Encyclopaedia of Mathematical Sciences

Volume 37

Editor-in-Chief: R.V. Gamkrelidze

A.I. Kostrikin I.R. Shafarevich (Eds.)

Algebra IV

Infinite Groups. Linear Groups

With 9 Figures

Springer-Verlag
Berlin Heidelberg New York
London Paris Tokyo
Hong Kong Barcelona
Budapest

Consulting Editors of the Series:
A.A. Agrachev, A.A. Gonchar, E.F. Mishchenko,
N.M. Ostianu, V.P. Sakharova, A.B. Zhishchenko

Title of the Russian edition:
Itogi nauki i tekhniki, Sovremennye problemy matematiki,
Fundamental'nye napravleniya, Vol. 37, Algebra 4
Publisher VINITI, Moscow 1989

Mathematics Subject Classification (1991):
20B07, 20Exx, 20Fxx, 20Gxx

ISBN 3-540-53372-9 Springer-Verlag Berlin Heidelberg New York
ISBN 0-387-53372-9 Springer-Verlag New York Berlin Heidelberg

Library of Congress Cataloging-in-Publication Data
Algebra 4. English. Algebra IV:
infinite groups, linear groups: with 9 figures
A.I. Kostrikin, I.R. Shafarevich (eds.).
p. cm.—(Encyclopaedia of mathematical sciences; v. 37)
Originally published as v. 37 of the serial: Itogi nauki i tekhniki.
Seriiā sovremennye problemy matematiki. Fundamental'nye napravleniiā
Includes bibliographical references and indexes.
ISBN 0-387-53372-9
1. Infinite groups. 2. Linear algebraic groups. I. Kostrikin, A. I. (Alekseĭ Ivanovich)
II. Shafarevich, I. R. (Igor Rostislavovich), 1923–. III. Title. IV. Series.
QA171.A5213 1993 512'.2—dc20 92-34231

Printed in the United States of America
Typesetting: Asco Trade Typesetting Ltd., Hong Kong
41/3140-543210—Printed on acid-free paper

List of Editors, Authors and Translators

Editor-in-Chief

R.V. Gamkrelidze, Russian Academy of Sciences, Steklov Mathematical Institute, ul. Vavilova 42, 117966 Moscow, Institute for Scientific Information (VINITI), ul. Usievicha 20a, 125219 Moscow, Russia

Consulting Editors

A.I. Kostrikin, I.R. Shafarevich, Steklov Mathematical Institute, ul. Vavilova 42, 117966 Moscow, Russia

Authors

A.Yu. Ol'shanskij, Department of Mathematics, Moscow State University, Leninskie Gory, 119899 Moscow, Russia
A.L. Shmel'kin, Department of Mathematics, Moscow State University, Leninskie Gory, 119899 Moscow, Russia
A.E. Zalesskij, Institute of Mathematics of the Belorussian Academy of Sciences, 220604 Minsk, Belorussia

Translator

J. Wiegold, School of Mathematics, University of Wales, Senghennydd Road, Cardiff CF2 4YH, United Kingdom

Contents

I. Infinite Groups

A.Yu. Ol'shanskij, A.L. Shmel'kin

Translated from the Russian
by J. Wiegold

Contents

Introduction

The achievements of Galois theory stimulated intensive study of permutation groups, and indeed in the early stages of its development, group theory was preoccupied almost exclusively with finite groups. However, under the influence of geometry, topology, and the theory of differential equations, there arose a pressing need to consider infinite groups of transformations. For example, Klein proposed that the classification of geometries should be linked with the description of the corresponding transformation groups. Parametric groups made their appearance in the works of Lie. Poincaré established the first contacts between combinatorial topology and group theory. Fedorov discovered a remarkable application of groups to the geometry of crystals.

There is an idea that has been developing over many years, and is by now completely obvious. It is this: whenever symmetry of a mathematical or physical object has a significant role to play – be the object of an algebraic nature, a differential equation, a crystallographic lattice or a geometrical invariant – it must have associated with it a group G of transformations preserving it (things like rigid motions, changes of variables, permutations of indices and so on). Groups thus emerged as measures of symmetry, and became indispensible tools for classifying the symmetries of a system.

The superposition of transformations in G is also in G (that is, it is a symmetry), and the term "group" came to be used for any set G with a law of composition defined on it: $(x, y) \mapsto xy \in G$, such that: 1) $(xy)z = x(yz)$, that is, the "product" is associative; 2) there is an element e in G such that $xe = ex = x$ for every x in G (e is called the identity element of G; for transformation groups, e is the identity mapping); 3) for every x in G, there exists an element x^{-1} in G (called the inverse of x) such that $xx^{-1} = x^{-1}x = e$. The study of group multiplication without restriction on the nature of the elements of G, and without the requirement that G be finite, was promoted in Shmidt's book entitled "Abstract Group Theory" and published in 1916. Groups are among the very first examples of abstract systems; at the beginning of this century, many other mathematical disciplines were re-examined on the group-theoretical model.

The breadth of the group axioms links groups with widely different branches of mathematics and natural science, and it causes new directions in the general theory to be closely ramified with the specifics inherent in them. The theory of topological groups, and other branches of mathematics where structures additional to the group structure are important, are outside the scope of our survey: their place is in specialist books on the relevant themes. Having once noted the various interconnections and motivations, we shall, in the main, be considering only properties of an abstract algebraical character, properties that do not depend on any analytical structure nor on the way transformation groups are represented.

Independently of the concrete nature of the elements, groups often appear together with natural generators and relations among them. The study of

groups given in this way is the subject of combinatorial group theory. Although the combinatorial and structural theories cannot be divorced from each other (the object of investigations is the same, even if the methods differ), we have separated this survey into two chapters, for the sake of convenience.

With very few exceptions, no special preparation is required of the reader, if such common mathematical terms as "normal subgroup", "factor-group", "simple group", "Cartesian product of groups" are discounted. Some of the conventional notation we use is as follows: S_n is the group of all permutations on $\{1, 2, \ldots, n\}$, $\mathrm{GL}_n(k)$ is the multiplicative group of all $n \times n$ non-singular matrices with coefficients in k (here $k = \mathbb{Z}$, the ring of integers, or a field), $\mathbb{Z}_n$ is the group (or ring) of residues modulo n: $\mathbb{Z}_n = \mathbb{Z}/n\mathbb{Z}$. A list of terms is to be found in the subject index.

The authors have concentrated their attention on the principal examples of groups, and these have been selected because of their connection with adjacent disciplines such as topology, linear and homological algebra, and other sciences. However, it should be noted that the axiomatic reconstruction of the foundations has enabled group theory to disover its independence, and has called to life a large collection of self-standing research programmes. For example, the study of groups with various "finiteness" conditions has given rise to the formulation of problems that are typical for the investigation of infinite groups, as well as to the creation of new methods. The emergence of the strong internal springs of the theory has inevitably meant that some parts of it are no longer in intimate contact with their origins, namely the most important examples of groups and types of groups.

At the present time, it is impossible to descry sufficient grounds for enunciating an aim such as that of describing up to isomorphism all (or even of wide classes of) groups, and indeed it is possible that the coming years will see the relation between the abstract and the concrete in group theory changing once again to the advantage of the latter. Mathematical logic, topological manifolds, the study of automorphisms of metric spaces and infinite graphs, algebraic geometry, and perhaps contemporary physics, all lead to the formulation of new problems in group theory. The concept of group, just like that of transformation or symmetry, has established itself firmly as one of the most fundamental of all mathematical ideas.

The aim of the present survey is to acquaint mathematicians in contiguous disciplines with the most important achievements of group theory and the methods employed there. Results occur either in the most up-to-date version, or in a form that seems most natural to the authors. In a survey like this, it is not possible to trace the historical development of this or that idea, nor of group theory in general. Individual remarks simply indicate the *dramatis personae*, not their contribution in group theory.

Chapter 1
Combinatorial Group Theory

By this we understand the large section of group theory that studies groups given in terms of generators and relations. In fact, that is how fundamental groups of topological spaces are usually described, and it was this fact that provided such a powerful impetus for the investigation of infinite groups. The origin of the combinatorial approach appeared in the works of Poincaré, Dehn, and Nielsen. The fruitfulness of the method is explained firstly by its universal nature – every group can be defined by generators and relations. Secondly, the combinatorial analysis of words and relations among them, as well as the "free constructions" typical of this theory, are often associated with natural geometrical representations and constructions.

In recent decades, combinatorial group theory has been influenced very strongly by mathematical logic: algorithm theory and model theory. The first of these influences to appear was the theorem of Novikov asserting the existence of finitely presented groups with algorithmically insoluble word problem.

§1. Free Groups

The involutions $a = (1, 2)$ and $b = (1, 3)$ in the symmetric group S_3 satisfy the relations $a^2 = 1$, $b^2 = 1$, $(ab)^3 = 1$, where 1 is the identity permutation. These relations are, of course, interpreted as being different, but in order to distinguish formally between their left-hand sides, it is necessary to study not only the (finite) group S_3, but also the infinite group consisting of words in the alphabet a, b, a^{-1}, b^{-1}.

1.1. Definition of Free Group. Let X be some finite or infinite alphabet. By a (group) *word* in X we understand an expression w of the form

$$w \equiv x_{i_1}^{\varepsilon_1} x_{i_2}^{\varepsilon_2} \dots x_{i_n}^{\varepsilon_n}, \tag{1}$$

where the x_{i_k} are in X, $\varepsilon_k = \pm 1$ for $k = 1, 2, \dots, n$, and $\equiv$ means letter-for-letter equality. The number n is called the *length* of w: $|w| = n$. The *empty* word has zero length, and it is denoted by 1. A word is said to be *uncancellable* if it contains no subwords of the form $x_j x_j^{-1}$ or $x_j^{-1} x_j$. The deletion of such a pair, that is, the replacement of such words of length 2 by the empty subword, is termed a cancellation. It is not difficult to check that every word can be reduced after finitely many cancellations to *uncancellable* form independent of the process of cancellation.

The *free group* $F(X)$ consists of all uncancellable words in the alphabet X, the product of words v and w being the result of bringing vw to uncancellable form. It is easy to see that this multiplication does in fact satisfy the group axioms, that the identity of $F(X)$ is the empty word, and the inverse of (1) is $w^{-1} \equiv x_{i_n}^{-\varepsilon_n} \dots x_{i_1}^{-\varepsilon_1}$. An equivalent definition of $F(X)$ gives its elements as classes of words having identical uncancellable form. With it, one does not need to worry about the constant reduction to uncancellable form of words appearing in any intermediate calculations in $F(X)$. The cardinality of $F(X)$ is called the *rank* of $F(X)$.

Free groups are torsion-free: every non-identity element w in $F(X)$ has infinite order, that is, $w^m \neq 1$ for every natural number m. Moreover, the equality $v^m = w^m$ in $F(X)$ implies that $v = w$ (that is, free groups are R-groups, groups with unique extraction of roots). The first property is obtained at once on representing w in the form $w = aba^{-1}$, where b is cyclically reduced, that is, its first and last letters are not mutually inverse. Then, w^m has uncancellable form $ab^m a^{-1}$, and this means that $w^m \neq 1$. The second property also follows from the definition and some simple combinatorial arguments. Equally easily, it can be shown that commuting elements v and w of $F(X)$ (that is, such that $vw = wv$ in $F(X)$) must be powers of one and the same element, in other words, must lie in one and the same cyclic subgroup.

Free groups have the following universal property. If G is any group and α any map $X \to G$, there exists a unique homomorphism $\bar{\alpha}$: $F(X) \to G$ extending α to $F(X)$ (if $w = x_{i_1}^{\varepsilon_1} \dots x_{i_n}^{\varepsilon_n}$, one simply writes $\bar{\alpha}(w) = \alpha(x_{i_1})^{\varepsilon_1} \dots \alpha(x_{i_n})^{\varepsilon_n}$). This property characterises free groups, in fact: if X is a subset of a group F which is universal in the sense indicated, then F is isomorphic with $F(X)$ in a natural way. The subset X is said to be a *basis* of F, and every group isomorphic to $F(X)$ is said to be free, that is, every group with a basis is free. The rank of a free group is an invariant: all its bases have the same cardinality.

1.2. Subgroups of Free Groups. As is clear from the definition given, a subset X of a group is a basis of a free subgroup F of G if and only if all products of the form (1) in which $x_{i_k}^{\varepsilon_k} x_{i_{k+1}}^{\varepsilon_{k+1}} \neq 1$ in G for all $k = 1, \dots, n-1$ are not the identity. For example, it is easy to show that the subset

$$Y = \{y_i \equiv x_1^i x_2 x_1^{-i} | i = 0, \pm 1, \dots\}$$

of the free group $F_2 = F(x_1, x_2)$ is a free basis for a subgroup H of F_2, namely the kernel of the homomorphism from F_2 onto the infinite cyclic group $F(x_1)$ such that $x_1 \mapsto x_1$, $x_2 \mapsto 1$. In particular, it follows from this that a free group of rank 2 can have a free subgroup of countably infinite rank.

This example is not an isolated one, since we have:

Theorem. *Every subgroup H of a free group F is a free group.*

The existence of bases for subgroups of groups of finite free rank was proved by Nielsen in 1921; Schreier did it in 1927 without restriction on the rank of F.

To describe Schreier's method, we recall the following definition.

Definition. A subset $M = \{a_i | i \in I\}$ of elements of a group G is said to be a set of *generators* if M is not contained in any proper subgroup of G. In other words, every element g in G is a product of the form $a_{i_1}^{\varepsilon_1} \dots a_{i_n}^{\varepsilon_n}$, where $i_k \in I$, $\varepsilon_k = \pm 1$. We write $G = \langle a_i | i \in I \rangle$ to denote this.

If H is a subgroup of a group G generated by a set M, a *generating* set for H can be chosen in the following way. Write G as a disjoint union of coset modulo H: $G = \bigcup_\alpha Hs_\alpha$. Having fixed the representatives s_α of the cosets modulo H, we define the function $g \mapsto \bar{g}$, which associates with each element g the representative of the coset containing g; that is, $Hg = H\bar{g}$ and $\bar{g} = s_\alpha$ for some s_α. It is convenient to take $\bar{g} = 1$ when g is in H. In this situation, the products $s_\alpha a_i \overline{s_\alpha a_i}^{-1}$ lie in H and generate H. In fact, we have first that $Hs_\alpha a_i = H\overline{s_\alpha a_i}$, so that $s_\alpha a_i \overline{s_\alpha a_i}^{-1}$ is in H. Secondly, $\overline{\bar{g}_1 g_2} = \overline{g_1 g_2}$ in all cases, since $H\bar{g}_1 g_2 = Hg_1 g_2$. Thus, we have for $g = a_1^{\varepsilon_1} \dots a_n^{\varepsilon_n} \in H$:

$$g = (1 \cdot a_1^{\varepsilon_1} \overline{1 \cdot a_1^{\varepsilon_1}}^{-1})(\overline{a_1^{\varepsilon_1}} a_2^{\varepsilon_2} \overline{a_1^{\varepsilon_1} a_2^{\varepsilon_2}}^{-1}) \dots (\overline{a_1^{\varepsilon_1} \dots a_{n-1}^{\varepsilon_{n-1}}} \cdot a_n \cdot \overline{a_1^{\varepsilon_1} \dots a_n^{\varepsilon_n}}^{-1}),$$

since $\overline{a_1^{\varepsilon_1} \dots a_n^{\varepsilon_n}} = \bar{g} = 1$. All we need is that

$$s_\alpha a_i^{-1} \overline{s_\alpha a_i^{-1}}^{-1} = (\overline{s_\alpha a_i^{-1}} \cdot a_i \cdot \overline{\overline{s_\alpha a_i^{-1}} \cdot a_i}^{-1})^{-1},$$

since $\overline{\overline{s_\alpha a_i^{-1}} \cdot a_i} = \overline{s_\alpha a_i^{-1} a_i} = \overline{s_\alpha} = s_\alpha$.

In particular, suppose that G is finitely generated (that is, some generating set M for it is finite), and that the number $|G:H|$ of cosets of G modulo H is finite (this is the *index* of H in G). Then H is also finitely generated.

For a free group F with basis X, it can be shown by induction that there is a so-called *Schreier system of representatives* modulo H. It has the property that every initial segment of the uncancellable representation of each s_α in terms of the letters in X is itself a coset representative. In this situation, the non-identity products $s_\alpha x_i \cdot \overline{s_\alpha x_i}^{-1}$ constitute a basis for H, so that H is free. Schreier established that a subgroup H of rank j in a free group of rank n has rank $1 + j(n - 1)$, by counting the number of these products.

1.3. Nielsen's Method. Automorphisms of Free Groups. It was Nielsen who created the theory of *cancellation* in free groups, which is one of the main methods of studying them. For a finitely generated subgroup H of a free group $F(X)$, a so-called *N-reduced system* of generators can be extracted using three types of transformations:

N1. In a generating set $\{a_1, \dots, a_m\}$ for H, replace a_i by a_i^{-1} for $i \in \{1, \dots, m\}$.

N2. For $i, j \in \{1, \dots, m\}$, $i \neq j$, replace a_i by $a_i a_j$.

N3. Reduce the number of generators by deleting any a_i such that $a_i = 1$ in $F(X)$.

A finite number of applications of these elementary *Nielsen transformations* reduces the system $\{a_i\}_{i=1}^n$ to an N-reduced system $\{b_j\}_{j=1}^l$, that is, a system such that:

1) $b_i \neq 1$ in $F(X)$ for $i = 1, \dots, l$;

2) if $v \equiv b_i^\varepsilon b_j^\delta$, where $\varepsilon, \delta = \pm 1$, and $\varepsilon \neq \delta$ if $i = j$, then $|v'| \geqslant |b_i|, |b_j|$ (here v' denotes the uncancellable form of v);

3) if $b_i^\varepsilon b_j^\delta \neq 1$ and $b_j^\varepsilon b_k^\xi \neq 1$ in $F(X)$, then

$$|(b_i^\varepsilon b_j^\delta b_k^\xi)'| > |b_i| - |b_j| + |b_k| \qquad (\varepsilon, \delta, \xi = \pm 1).$$

Theorem (Nielsen). *Every N-reduced system of uncancellable words is a basis of a subgroup H for which the sum of the lengths of the words contained in it with respect to the alphabet X is the smallest for all generating sets of H.*

When $X = \{x_1, \dots, x_n\}$ and $H = F(X)$, it follows from this that every generating set for $F(X)$ can be carried into $\{x_1, \dots, x_n\}$ by a finite number of elementary Nielsen transformations. Thus, the elementary automorphisms, that is, those given by the elementary Nielsen transformations, generate the automorphism group Aut $F(X)$. Aut $F(X)$ is finitely presented (see section 2.1). A set of defining relatons is to be found in Magnus, Karras and Solitar (1966).

Birman has established the following criterion in terms of free differential calculus (the necessary definitions are given in section 4.3).

Theorem. *A set of words $\{w_1, \dots, w_n\}$ in the alphabet $X = \{x_1, \dots, x_n\}$ is a basis for $F(X)$ if and only if the Fox matrix $\left(\frac{\partial w_i}{\partial x_j}\right)$ is invertible in the group ring $\mathbb{Z}[F(X)]$.*

Every Nielsen automorphism fixes almost all elements in the basis, so that Aut $F(X)$ is not generated by the Nielsen automorphisms when X is infinite. However, the subgroup $\mathrm{Aut}_N\, F(X)$ they generate is "dense" in Aut $F(X)$:

Theorem. *For every finite set of elements $w_1, \dots, w_n$ in $F(X)$ and every α in* Aut $F(X)$, *there is an automorphism β in* $\mathrm{Aut}_N\, F(X)$ *such that $\alpha(w_i) = \beta(w_i)$ for each $i = 1, \dots, n$.*

The free group $F_n = F(x_1, \dots, x_n)$ has a natural homomorphism onto the free abelian group of rank n (see section 1 of Chapter 2). Thus, there is a canonical homomorphism ϕ of Aut F_n into $\mathrm{GL}_n(\mathbb{Z})$. Using the fact that $\mathrm{GL}_n(\mathbb{Z})$ is generated by elementary matrices, it is easy to see that ϕ is surjective, so that it is an epimorphism. Clearly, its kernel contains all the inner automorphisms (that is, automorphisms of the form α_a, $a \in F_n$, such that $\alpha_a(g) = aga^{-1}$ for all g in F_n). For $n = 2$, Nielsen proved the following result:

Theorem. *The kernel of the homomorphism* ϕ: Aut $F_2 \to \mathrm{GL}_2(\mathbb{Z})$ *consists exactly of the inner automorphisms of F_2.*

The analogous assertion for $n \geqslant 3$ is false.

There exists an algorithm allowing one to decide, for a pair u, v of words in $F(X)$, whether or not there is an automorphism carrying the one to the other. Rather than dwelling on particular results about the automorphism group of $F(X)$, we state next a powerful assertion, proved quite recently by Gersten. Unfortunately, an explanation of the proof would require the introduction of a large number of special terms.

Theorem. *The fixed-point subgroup of every automorphism of a free group of finite rank is of finite rank.*

Howson's theorem (see section 1.4) shows that this result can be extended to the fixed-point subgroup of any finitely generated subgroup G of Aut F_n.

1.4. Other Realisations of Free Groups. In addition to the representation of a free group as a group of words, there are some other useful descriptions. For example, the *Magnus representation* of $F(X)$ in the ring of formal power series in non-commuting variables has been a particularly fruitful one. The ring $A(\mathbb{Z}, X)$ consists of the formal sums of the form $\sum_{i=0}^{\infty} u_i$, where u_i is a homogeneous polynomial of degree i, that is, a finite integral linear combination of "non-commuting" monomials of the form

$$x_{i_1}^{l_1} \dots x_{i_n}^{l_n}, \quad \text{where} \quad i_k \neq i_{k+1} \quad \text{for} \quad k = 1, \dots, n-1, \quad l_k > 0, \quad \sum_{k=1}^{n} l_k = i.$$

The multiplication of monomials is simple juxtaposition, and this operation extends naturally to series. The set of invertible elements of $A(\mathbb{Z}, X)$ consists of the series with free term ± 1. For example, $(1 + x_j)^{-1} = 1 - x_j + x_j^2 - x_j^3 + \cdots$.

Lemma. *The elements* $1 + x_j$ *with* $x_j \in X$ *freely generate a subgroup of the multiplicative group isomorphic with* $F(X)$.

For the proof, we need to check that the series

$$(1 + x_{i_1})^{l_1} \dots (1 + x_{i_n})^{i_n},$$

where $n \geqslant 1$, $i_k \neq i_{k+1}$, $l_k \neq 0$, is different from 1. It can be rewritten in the form

$$(1 + l_1 x_{i_1} + \cdots) \dots (1 + l_n x_{i_n} + \cdots),$$

whence it is clear that the monomial $x_{i_1} \dots x_{i_n}$ occurs in the product with non-zero coefficient $l_1 \dots l_n$. This proves the lemma.

This description of $F(X)$ enabled Magnus to study the lower central series (see section 4.1 in Chapter 2) of a free group. It turned out that $\gamma_m(F(X))$ consists of the series $\sum_{i=0}^{\infty} u_i$ in $F(X)$ for which $u_1 = \cdots = u_{m-1} = 0$. The next theorem follows immediately from this.

Theorem. 1) $\bigcap_{m=1}^{\infty} \gamma_m(F(X)) = \{1\}$; 2) *the nilpotent factorgroups* $F(X)/\gamma_m(F(X))$ *are torsion-free*; 3) *the factors* $\gamma_m(F(X))/\gamma_{m+1}(F(X))$ *are free abelian.*

Witt discovered the *number of generators* of these abelian groups; if $F = F_n$ is free of rank n, the rank of $\gamma_m(F_n)/\gamma_{m+1}(F_n)$ as free abelian group (see section 1 of Chapter 2) is

$$\frac{1}{m} \sum_{d|m} \mu(d) n^{m/d},$$

where μ is the Möbius function.

Sanov found a beautiful linear representation of F_2, in 1947.

Theorem. *The correspondence* $x_1 \mapsto \begin{pmatrix} 1 & 2 \\ 0 & 1 \end{pmatrix}$, $x_2 \mapsto \begin{pmatrix} 1 & 0 \\ 2 & 1 \end{pmatrix}$ *defines an isomorphic embedding of* F_2 *into* $\mathrm{GL}_2(\mathbb{Z})$.

If we change the matrices to

$$A_\mu = \begin{pmatrix} 1 & \mu \\ 0 & 1 \end{pmatrix} \quad \text{and} \quad B_\mu = \begin{pmatrix} 1 & 0 \\ \mu & 1 \end{pmatrix},$$

the corresponding representation is an isomorphism for all μ with $|\mu| \geqslant 2$. Clearly, all transcendentals μ are also "free". Several articles (see Lyndon and Schupp (1977)) are devoted to determining which complex numbers are free and which not, in the sense indicated. Interest in this problem has been stimulated by the *Tits alternative*: a finitely generated linear group either contains a free subgroup of rank 2, or else it has a soluble subgroup of finite index (see Part II, "Linear Groups", of this volume).

Residual finiteness of free groups is easily deduced from the Sanov representation. At this point, we give a more general definition.

Definition. Let $\mathscr{K}$ be a class of groups. A group G is said to be residually in $\mathscr{K}$ if, for every $g \in G \setminus \{1\}$, there exists an epimorphism ϕ of G onto a group K in $\mathscr{K}$ such that $\phi(g) \neq 1$.

A group G is residually in $\mathscr{K}$ if and only if it is isomorphic with a subgroup of a Cartesian product of groups from $\mathscr{K}$ such that the projection onto each component is an epimorphism.

If $\mathscr{K}$ is the class of all finite groups, the groups that are residually in $\mathscr{K}$ are called *residually finite*. A group G is residually finite if (and only if) the intersection of all normal subgroups of finite index is the identity subgroup.

To prove that the free group $F(X)$ is residually finite, it is enough to do it when $|X| < \infty$, since every word w in $F(X)$ can be written in terms of a finite subset of the alphabet, and maps to a nonempty word under an epimorphism to some F_r. The example in section 1.2 shows that F_r is embeddable in F_2, so that by Sanov's theorem it is embeddable in the group $SL_2(\mathbb{Z})$ of integer matrices of determinant 1. Finally, this last group is residually in the set of finite groups of the form $SL_2(\mathbb{Z}_n)$, the groups of unimodular matrices over the residue class rings $\mathbb{Z}_n$, $n = 1, 2, \ldots$.

Definition. A group G is said to be *Hopfian* if every epimorphism $G \to G$ is an automorphism of G (that is, it has trivial kernel).

Every finitely generated residually finite group is Hopfian (see Magnus, Karrass and Solitar (1966)). Thus, the free group F_n is Hopfian, a fact that can also be deduced from the theorem of Magnus in 1.3.

The following assertion of Burns about free groups is significantly stronger than residually finiteness. (See section 2.2 for the definition of free product.)

Theorem. *Let A be a finite subset of a free group F, and H a finitely generated subgroup of F such that $H \cap A = \varnothing$. Then H is a free factor of a subgroup K of finite index in F such that $K \cap A = \varnothing$.*

It is easy to deduce from this that, if H is a finitely generated subgroup of F containing a non-trivial normal subgroup, then $|F:H| < \infty$. In particular, non-trivial finitely generated normal subgroups of free groups are of finite index. Howson's theorem can also be deduced from that of Burns:

Theorem. *The intersection of two finitely generated subgroups of a free group is finitely generated.*

§ 2. Defining Relations and Free Constructions

2.1. Presentations of Groups. Suppose that a set $M = \{a_i | i \in I\}$ of generators is chosen in a group G. As we saw in § 1, there is just one epimorphism of the free group $F(X)$ on the alphabet $X = \{x_i | i \in I\}$ onto G extending the map $x_i \mapsto a_i$. The epimorphism ϕ is called a *presentation* of G. It depends on the generating set for G. The elements of Ker ϕ are called relations in G (among the generators in M); however, if $x_{i_1}^{\varepsilon_1} \dots x_{i_n}^{\varepsilon_n} \in \operatorname{Ker} \phi$, it is usual to write the relation in the form $a_{i_1}^{\varepsilon_1} \dots a_{i_n}^{\varepsilon_n} = 1$ (this is, of course, a true equation in G). The expression $a_{i_1}^{\varepsilon_1} \dots a_{i_k}^{\varepsilon_k} = a_{i_n}^{-\varepsilon_n} \dots a_{i_{k+1}}^{-\varepsilon_{k+1}}$ is also permitted.

If the set R of relations is not contained in any normal subgroup N of $F(X)$ such that $N \subsetneqq \operatorname{Ker} \phi$, then R is called a *set of defining relations* for G (among the generators in M). Since $G \cong F(X)/\operatorname{Ker} \phi$, G is uniquely defined (up to isomorphism) by the pair X, R; it is suggested in Kargapolov and Merzlyakov (1982) that the pair be called a genetic code for G. The expression $G = \langle M; R \rangle$ is normally used to express the fact that the group is given by generators and relations. If all the relations in R hold for elements b_i $(i \in I)$ of some group H, the map $a_i \mapsto b_i$ extends to a homomorphism $G \to H$. This property is sometimes called von Dyck's theorem; defining relations first made their appearance in a paper published by von Dyck in 1882/83.

A typical situation in combinatorial group theory is when one has a presentation, that is, X and R are the initial data; often this group is not given in any other way (say, as a transformation group), and its properties are investigated by analyzing the consequences of the defining relations.

Definition. The group G is said to be *finitely presented* if M and R are finite in some presentation $G = \langle M; R \rangle$.

How can one prove that some set of relations in a known group are defining relations? To do this, it must be shown that every relation r lies in the normal closure of R in $F(X)$, that is, in the least normal subgroup of $F(X)$ containing R. This holds if and only if r can be written in the form

$$r = \prod u_k r_k u_k^{-1},$$

where $r_k \in R$, $u_k \in F(X)$. In other words, it must be shown that r is 1 in any group where all the words in R are 1.

We consider there the example provided at the beginning of the chapter. If $r = 1$ is a relation in S_3, then using the relations $a^2 = 1$ and $b^2 = 1$ in R, we can bring r to the form $(a)bab\dots aba(b)$ where the letters in brackets may be absent. The length of this expression can to be taken to be less than 6, as is easily seen using the relation $(ab)^3 = 1$. From this it follows that r is empty, since it is simple to check that the products , ab, aba, $abab$, $ababa$ are not 1 in S_3. It has thus been shown that $r = 1$ is a consequence of the three relations, that is,

$$S_3 = \langle a, b; a^2 = b^2 = (ab)^3 = 1\rangle.$$

It is even easier to show that the free abelian group of rank n (see § 1 of Chapter 2) can be given by the presentation

$$\langle a_1, \dots, a_n; a_i a_j a_i^{-1} a_j^{-1} = 1, 1 \leqslant i < j \leqslant n\rangle.$$

A useful method for determining sets of defining relations for finite groups, as well as proving that a group given by defining relations is actually finite, is the method of enumerating cosets (see Kargopolov and Merzlyakov (1982)). It is very convenient for implementation on a computer, and gives a description of the group in terms of permutations.

There is considerable arbitrariness in presenting groups in terms of generators and relations, both in the choice of generators and that of defining relations. However, any presentation can be taken into any other using *Tietze transformations*. For finitely presented groups, they are of the following type:

1) if $r = 1$ is a consequence of the relations in R, add r to R;

2) if $r \in R$, and the relation $r = 1$ can be deduced from the remaining relations in R, delete r from R;

3) if W is a word in the alphabet X, adjoin a letter w to the alphabet and add the relation $w = W$ to the set of defining relations;

4) if a defining relation has the form $w = W$, where the word W is an expression in the letters other than w, delete w from X, delete $w = W$ from the list of defining relations, and replace w by W in the remaining defining relations.

For example, in the generators $a = (1, 2)$, $c = (1, 2, 3)$, our group has presentation

$$\langle a, c; a^2 = 1, c^3 = 1, aca^{-1} = c^{-1}\rangle.$$

It is obtained from the example considered above using the following Tietze transformations:

$$\begin{aligned}
&\langle a, b; a^2 = 1, b^2 = 1, (ab)^3 = 1\rangle \\
&\overset{(3)}{\to} \langle a, b, c; a^2 = 1, b^2 = 1, (ab)^3 = 1, c = ba\rangle \\
&\overset{(1)}{\to} \langle a, b, c; a^2 = 1, b^2 = 1, (ab)^3 = 1, c = ba, b = ca^{-1}\rangle \\
&\overset{(4)}{\to} \langle a, c; a^2 = 1, (ca^{-1})^2 = 1, (aca^{-1})^3 = 1, c = caa^{-1}\rangle \\
&\overset{(2)}{\to} \langle a, c; a^2 = 1, (ca^{-1})^2 = 1, (aca^{-1})^3 = 1\rangle \\
&\overset{(1)}{\to} \langle a, c; a^2 = 1, (ca^{-1})^2 = 1, (aca^{-1})^3 = 1, c^3 = 1\rangle
\end{aligned}$$

$$\overset{(2)}{\rightarrow} \langle a, c; a^2 = 1, c^3 = 1, (ca^{-1})^2 = 1\rangle$$
$$\overset{(1)}{\rightarrow} \langle a, c; a^2 = 1, c^3 = 1, (ca^{-1})^2 = 1, aca^{-1} = c^{-1}\rangle$$
$$\overset{(2)}{\rightarrow} \langle a, c; a^2 = 1, c^3 = 1, aca^{-1} = c^{-1}\rangle.$$

Tietze's theorem remains true for groups with infinitely many generators or relations. It is simply necessary to apply a single "elementary" transformation to add or delete infinitely many letters or relations all at once. Finally, we note that Tietze's theorem carries over without change to rings, semigroups and other algebraic systems.

A presentation can be found for a subgroup K of $G = \langle X; R\rangle$ using the *Reidemeister-Schreier rewriting process*. It comes to choosing a Schreier system of generators $s_\alpha x_i \overline{s_\alpha x_i}^{-1}$ (see section 1.2) for the subgroup H of $F(X)$ which is the complete inverse image of K under the homomorphism $F(X) \to G$; this corresponds to a new alphabet $Y = \{y_{\alpha,i}\}$. Further, words of the form $s_\alpha r s_\alpha^{-1}$ ($r \in R$) lie in H, and it is natural to compare them with the words in Y using the rewriting process, that is, with the expressions of the words $s_\alpha r s_\alpha^{-1}$ as products of elements $(s_\alpha x_i \overline{s_\alpha x_i}^{-1})^{\pm 1}$. The set U of all words obtained in this way constitutes a set of defining relations for H: $H = \langle Y; U\rangle$.

For example, consider

$$G = \langle a, b, c, d; aba^{-1}b^{-1}cdc^{-1}d^{-1} = 1\rangle,$$

the fundamental group of a two-sphere with two handles. The kernel of the homomorphism of G onto a cyclic group of order 2 such that $b \mapsto 1$, $c \mapsto 1$, $d \mapsto 1$ is of index 2 in G; $\{1, a\}$ is a Schreier system of representatives for the corresponding subgroup H, and

$$\{a^2, b, c, d, aba^{-1}, aca^{-1}, ada^{-1}\} \tag{2}$$

is a Schreier system of generators for H. In this case,

$$aba^{-1}b^{-1}cdc^{-1}d^{-1} = (aba^{-1})\cdot b^{-1}\cdot c\cdot d\cdot c^{-1}\cdot d^{-1}$$

and

$$a(aba^{-1}b^{-1}cdc^{-1}d^{-1})a^{-1}$$
$$= a^2\cdot b\cdot a^{-2}\cdot(aba^{-1})\cdot(aca^{-1})\cdot(ada^{-1})\cdot(aca^{-1})^{-1}\cdot(ada^{-1})^{-1}.$$

This means that in the alphabet $\{t, u, v, w, x, y, z\}$ corresponding to the set (2), we get

$$K = \langle t, u, v, w, x, y, z; xu^{-1}vwv^{-1}w^{-1} = 1, tut^{-1}xyzy^{-1}z^{-1} = 1\rangle.$$

Removing x by a Tietze transformation using the first relation, we get

$$K = \langle t, u, v, w, y, z; tut^{-1}wvw^{-1}v^{-1}uyzy^{-1}z^{-1} = 1\rangle.$$

It is easy to see, further, that Tietze transformations enable us to introduce letters $w_1 = uwu^{-1}$ and $v_1 = uvu^{-1}$ replacing w and u; having done that, we see that

$$K = \langle t, u, v_1, w_1, y, z; tut^{-1}u^{-1}w_1v_1w_1^{-1}v_1^{-1}yzy^{-1}z^{-1}\rangle$$

is in fact the fundamental group of a sphere with three handles.

2.2. Free Constructions. The free product, like the direct product, has a natural analogue in topology. The fundamental group of the Cartesian product of two path-connected spaces X_1 and X_2 is the direct product $G_1 \times G_2$. The fundamental group of the union of X_1 and X_2 joined at a single point is isomorphic to the free product $G_1 * G_2$.

Definition. The *free product of the groups* G_i, $i \in I$, is a group $G = *_{i \in I} G_i$, together with homomorphisms ϕ_i: $G_i \to G$ such that, for every group A and all homomorphisms α_i: $G_i \to A$, there exists just one homomorphism ϕ: $G \to A$ such $\phi\phi_i = \alpha_i$ for every i in I.

Such a group G exists, and is unique up to a natural isomorphism. If the G_i are give by defining relatons in disjoint generating sets, a presentation for G is obtained by uniting the generating sets and the sets of defining relations of the G_i. The homomorphisms ϕ_i are injective (that is, they are monomorphisms), and the G_i are usually identified with subgroups of G. Every non-identity element g in G has a canonical representation of the form $g = g_1 g_2 \dots g_n$, where g_k and g_{k+1} are non-identity elements from different factors for $k = 1, \dots, n-1$; n is the length of g.

Every free group $F(X)$ decomposes as the free product of the infinite cyclic subgroups $\langle x_i \rangle$, $x_i \in X$. The main structure theorem about free products is the Kurosh subgroup theorem:

Theorem. *Every subgroup H of a free product $G = *_{i \in I} G_i$ decomposes as a free product of subgroups, each of which is either conjugate*[1] *to a subgroup of one of the factors G_i, or is infinite cyclic.*

Suppose that $G = G_1 * \dots * G_n$ and that G is generated by a finite set $M = \{g_1, \dots, g_m\}$. Grushko's theorem says that Nielsen transformations can be applied to M to give a set

$$M' = \{g_{11}, \dots, g_{1,m_1}; \dots; g_{n1}, \dots, g_{n,m_n}\}$$

such that $\langle g_{i1}, \dots, g_{in_i} \rangle = G_i$. In particular, the minimum number of generators of G is the sum of the minimum numbers of generators of the factors G_i.

Definition. Let A_1 and A_2 be isomorphic subgroups of groups G_1 and G_2 respectively with presentations $G_1 = \langle X_1, R_1 \rangle$, $G_2 = \langle X_2, R_2 \rangle$. By the free product of G_1 and G_2 with subgroups A_1 and A_2 amalgamated according to the isomorphism ϕ: $A_1 \to A_2$ we mean the group $G = \langle G_1 * G_2, A_1 = A_2, \phi \rangle$ with

[1] We recall that an element of the form aga^{-1} (subgroup of the form aHa^{-1}) is said to be conjugate to the element g (to the subgroup H) in G.

generating set the disjoint union $X_1 \cup X_2$ and defining relations the union of R_1, R_2 and the set $\{a = \phi(a) | a \in A_1\}$.

The construction of the free product with amalgamation over any number of factors is fully analogous. (A single subgroup A_i is chosen in each G_i!) It does not depend on the choice of the presentations of the factors. The groups G_i are isomorphically embedded in G and are identified with their images there, and $G_i \cap G_j = A_i = A_j = A$.

Every element g of G has a canonical expression obtained after choosing coset representatives: $G_i = \bigcup_\alpha A_i s_{i\alpha}$. Namely,

$$g = a s_{i_1\alpha_1} s_{i_2\alpha_2} \dots s_{i_n\alpha_n},$$

where the $s_{i_k\alpha_k}$ are not the identity, $i_k \neq i_{k+1}$ for $k = 1, \dots, n$, and $a \in A$.

The simplest example of a free product is the group $\mathrm{PSL}_2(\mathbb{Z})$ of fractional linear transformations of the upper complex half-plane:

$$z \mapsto \frac{az + b}{cz + d}, \tag{3}$$

where a, b, c, d are integers and $ad - bc = 1$. It is the free product of the cyclic subgroups $\langle\phi\rangle$ of order 3 and $\langle\psi\rangle$ of order 2, where $\phi(z) = 1 - \frac{1}{z}$, $\psi(z) = -\frac{1}{z}$ (see section 3.2). This group is a natural homomorphic image of the group $SL_2(\mathbb{Z})$ of unimodular matrices: $A \mapsto \phi$, $B \mapsto \psi$, where $A = \begin{pmatrix} 1 & -1 \\ 1 & 0 \end{pmatrix}$, $B = \begin{pmatrix} 0 & -1 \\ 1 & 0 \end{pmatrix}$ (generally, the matrix $\begin{pmatrix} a & b \\ c & d \end{pmatrix}$ corresponds to the transformation (3)). The kernel of this map consists exactly of the matrices $\pm E$. Thus

$$SL_2(Z) = \langle A, B; A^6 = 1, B^4 = 1, A^3 = B^2 \rangle$$

is the free product of cyclic groups of order six and four amalgamating subgroups of order 2.

The Seifert-van Kampen theorem points up the importance of studying free constructions in topology. If the space $Z = X \cup Y$ is the union of path-connected spaces X and Y with nonempty path-connected intersection V such that the natural homomorphisms $\pi_1(V) \to \pi_1(X)$ and $\pi_1(V) \to \pi_1(Y)$ of fundamental groups are monomorphisms, then $\pi_1(Z)$ is isomorphic with the free product of $\pi_1(X)$ and $\pi_1(Y)$ with $\pi_1(V)$ amalgamated.

There is another construction that is naturally studied alongside free products, namely the HNN-extensions (or Higman-Neumann-Neumann extensions).

Definition. Let G be a group, A and B subgroups of G, and ϕ: $A \to B$ an isomorphism. The HNN-*extension* G^* of G with respect to ϕ is the group obtained by adding a single letter t to the list of generators of G and the relations $tat^{-1} = \phi(a)$ for all a in A to the set of defining relations.

Once again we need to say that G is embedded naturally in G^* (G is called the base subgroup of G^*), and that the definition of G^* is invariant with respect to the choice of presentation for G.

A sequence $g_0, t^{\varepsilon_1}, g_1, \ldots, t^{\varepsilon_n}, g_n$, where $g_i \in G$, $\varepsilon_i = \pm 1$, is said to be reduced if the elements t, g_i, t^{-1} with g_i in A do not occur consecutively in it, nor t^{-1}, g_i, t with g_i in B. The foundation for the combinatorial study of HNN-extensions is the Novikov-Britton lemma:

Lemma. *If $g_0, t^{\varepsilon_1}, g_1, \ldots, t^{\varepsilon_n}, g_n$ is a reduced sequence with $n \geqslant 1$, then $g_0 t^{\varepsilon_1} g_1 \ldots t^{\varepsilon_n} g_n$ is different from 1 in G^*.*

HNN-extensions also have a topological significance. Let X and Y be open path-connected subspaces of a path-connected space Z for which there exists a homeomorphism $X \to Y$ such that the fundamental groups $\pi_1(X)$ and $\pi_1(Y)$ are isomorphically embedded in $\pi_1(Z)$. We construct a space W by attaching the space $X \times [0, 1]$ (as handle) to Z, identifying $X \times \{0\}$ with X and $X \times \{1\}$ with Y. The fundamental group $\pi_1(W)$ of W is the HNN-extension of $\pi_1(X)$ relative to the isomorphism between its subgroups $\pi_1(X)$ and $\pi_1(Y)$.

HNN-extensions first found group-theoretical application in the proof of an embedding theorem in a paper by Higman, Neumann and Neumann in 1949.

Theorem. *Every countable group G is isomorphic with a subgroup of a two-generator group A.*

If G is finitely presented, so is A. It was also proved in that article that every countable group can be embedded in a countably infinite group in which all elements of given order are conjugate. As Goryushkin has proved, every countable group can even be embedded in a 2-generator simple group. (This is just one of many examples which show that 2-generator groups are just as complicated as arbitrary finitely generated groups. The same is true in relation to many special classes of groups, to be considered in the second chapter). The first examples of infinite finitely presented simple groups were found by R.J. Thompson while investigating the automorphism groups of free Cantor algebras (in this context, see also the survey "Identities" in the present series).

§ 3. Properties of Free Constructions

3.1. Subgroups of Free Products with Amalgamation and HNN-Extensions. There is no natural analogue of the Kurosh subgroup theorem for free products in the case of free products with amalgamation nor HNN-extensions, when these constructions are looked at on their own. It has turned out that the answers to the two subgroup problems come only when these basic constructions are studied together. Structure theorems were obtained by Karrass and Solitar in 1970–71.

In order to state subgroup theorems, we shall first generalise the concept of free product with amalgamation. Let T be a finite or infinite tree, that is, a connected graph without cycles. Suppose further that each vertex v in the vertex-set V is associated with a group G_v, and that to each (non-oriented) edge e in E (E is the edge set) there is a subgroup H_e embedded in G_v and G_w *via* monomorphisms $\phi_{e,v}$ and $\phi_{e,w}$ if e joins v and w.

Definition. Suppose that the groups $G_v = \langle X_v; R_v \rangle$ are given on disjoint generating sets X_v. Their *tree product* is the group G with generating set $\bigcup_{v \in V} X_v$ and defining relations consisting of $\bigcup_{v \in V} R_v$ together with all relations of the form

$$\varphi_{e,v}(h) = \varphi_{e,w}(h) \qquad \text{for all} \qquad h \in H_e,\, e \in E.$$

Every subgroup H of the free product $G = \langle G_1 * G_2, A_1 = A_2, \phi \rangle$ with amalgamation can be described as follows. Let $\{s_\alpha\}$ be a system of representatives of the double cosets of G modulo (H, G_1), so that $G = \bigcup_\alpha H s_\alpha G_1$. Similarly, let $\{t_\beta\}$ be a system of representatives modulo (H, G_2). We set

$$H_\alpha = H \cap s_\alpha G_1 s_\alpha^{-1}, \qquad H_\beta = H \cap t_\beta G_2 t_\beta^{-1}.$$

Theorem. *The subgroups H_α and H_β generate in H their tree product K with amalgamated subgroups of the form $H \cap s_\alpha A_1 s_\alpha^{-1}$ and $H \cap t_\beta A_2 t_\beta^{-1}$, and H is an* HNN-*extension with base subgroup K.*

However, the conjugacy is introduced not simply for a single pair of subgroups of K, but rather for infinite sets of pair of isomorphic subgroups.

A parallel theorem holds for subgroups of HNN-extensions. Every subgroup H of an HNN-extension G^* with base subgroup G can be given as an HNN-extension of a tree product of subgroups of the form $H \cap aGa^{-1}$, $a \in G^*$. (See Lyndon and Schupp (1977) for more details.)

3.2. Free Constructions as Transformation Groups. Free products (with amalgamation) and HNN-extensions can be characterised as transformation groups, as was observed in papers of Macbeath and Maskit. The statement of our first theorem is suggested by a simple example. Let $G = G_1 * G_2$ be a free product acting regularly on itself by left translations: $x \to gx$. Further, suppose that S_1 (S_2 respectively) is the set of non-identity elements of G whose normal forms start with factors in G_1 (in G_2 respectively). Clearly, $gS_2 \subseteq S_1$ for $g \in G_1 \backslash \{1\}$ and $gS_1 \subseteq S_2$ for $g \in G_2 \backslash \{1\}$. The converse is also true.

Theorem. *If G is a group of permutations of some set S generated by subgroups G_1 and G_2 in such a way that there are two disjoint subsets S_1 and S_2 of S for which $gS_2 \subseteq S_1$ when $g \in G_1 \backslash \{1\}$ and $gS_1 \subseteq S_2$ when $g \in G_2 \backslash \{1\}$, then $G = G_1 * G_2$, except possibly when $|G_1| = |G_2| = 2$ and $G = \langle a, b; a^2 = b^2 = (ab)^n = 1 \rangle$ is dihedral of order $2n$.*

Using this theorem, we shall show that the transformations $\phi(z) = 1 - \frac{1}{z}$ and $\psi(z) = -\frac{1}{z}$ of the upper complex half-plane mentioned in 2.2 really do generate free factors, that is, $\mathrm{PSL}_2(\mathbb{Z}) = \langle\phi\rangle_3 * \langle\psi\rangle_2$. It is immediately obvious that ϕ^3 and ψ^2 are the identity. For S we take the upper half-plane, and for S_1 the first quadrant: $S_1 = \{z \mid \operatorname{Re} z > 0, \operatorname{Im} z > 0\}$. Similarly, S_2 is the second quadrant $\{z \mid \operatorname{Re} z < 0, \operatorname{Im} z > 0\}$. It is obvious that $\psi(S_1) \subseteq S_2$ and $\phi(S_2) \subseteq S_1$; since $\phi^2(z) = \dfrac{1}{1-z}$, we have $\phi^2(S_2) \subseteq S_1$. The requirements of the theorem are satisfied, so we just have to apply it!

Similarly, to show that the group $\langle A_\mu, B_\mu\rangle$ generated by the matrices $A_\mu = \begin{pmatrix} 1 & \mu \\ 0 & 1 \end{pmatrix}$ and $B = \begin{pmatrix} 1 & 0 \\ \mu & 1 \end{pmatrix}$ is free when $|\mu| \geqslant 2$ (see 1.4), it is enough to observe that it acts as a group of fractional linear transformations of the Riemann sphere $\mathbb{C} \cup \{\infty\}$ given by $A_\mu(z) = z + \mu$ and $B_\mu(z) = \dfrac{z}{\mu z + 1}$. Now define $S_1 = \{z \mid |z| > 1\}$ and $S_2 = \{z \mid |z| < 1\}$ and note the following conclusions, which are obvious for $|\mu| \geqslant 2$:

$$A_\mu^k(S_2) \subseteq S_1 \quad \text{and} \quad B_\mu^k(S_1) \subseteq S_2 \quad \text{for} \quad k = \pm 1, \pm 2, \ldots.$$

The dihedral group really is an exception to the theorem. It is enough to consider its natural representation as the group of symmetries of a dihedron consisting of regular pyramids with a common base (Fig. 1); a and b are chosen to be the rotations of three-space through an angle π about axes that are axes of symmetry of the base of the dihedron with constitutent angle $\dfrac{\pi}{n}$. We set $S_1 = \{o_1\}$, $S_2 = \{o_2\}$, where o_1 and o_2 are opposite vertices of the pyramids.

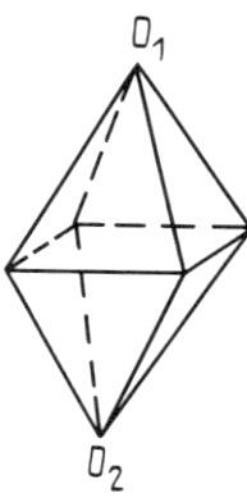

Fig. 1

We have the following characterisation for free products with amalgamation.

Theorem. *Let G be a group of permutations of a set S generated by subgroups G_1 and G_2 whose intersection H is a proper subgroup in each of them such that the indices $|G_1 : H|$ and $|G_2 : H|$ are not both 2. Suppose that there are disjoint subsets S_1 and S_2 of S such that $gS_2 \subseteq S_1$ for $g \in G_1 \setminus H$ and $gS_1 \subseteq S_2$ for $g \in G_2 \setminus H$, and also $hS_1 \subseteq S_1$, $hS_2 \subseteq S_2$ for $h \in H$. Then G is the free product of G_1 and G_2 with H amalgamated.*

We show now how this assertion can be used to obtain a proof of a theorem of Nagao.

Let F be a field and $F[t]$ the ring of polynomials in a single variable t over F. The group $\mathrm{GL}_2(F)$ of nonsingular matrices over F contains the subgroup $T(F)$ consisting of the upper triangular matrices; and $T(F[t])$ is the analogous subgroup of the group $\mathrm{GL}_2(F[t])$ of invertible matrices over $F[t]$. This group $\mathrm{GL}_2(F[t])$ acts by left multiplication on the set S consisting of the columns–pairs of polynomials. Let S_1 be the set consisting of nonzero pairs (f, g) such that $\deg f \leqslant \deg g$, and set $S_2 = S \setminus S_1$. It is clear that $T(F)S_1 \subseteq S_1$, $T(F)S_2 \subseteq S_2$, $gS_2 \subseteq S_1$ for $g \in \mathrm{GL}_2(F) \setminus T(F)$, and $gS_1 \subseteq S_2$ for $g \in T(F[t]) \setminus T(F)$. Consequently, $\mathrm{GL}_2(F[t])$ is the free product of $\mathrm{GL}_2(F)$ and $T(F[t])$ with $T(F)$ amalgamated. (We leave the proof that $\mathrm{GL}_2(F[t]) = \langle \mathrm{GL}_2(F), T(F[t]) \rangle$ to the reader.)

Finally, a transformation group G is an HNN-extension in the following situation. Let G^* be a group generated by a subgroup G and an element t, acting as a permutation group on a set S. Suppose further that A_1 is a subgroup of G such that the subgroup $A_{-1} = tA_1t^{-1}$ is also contained in G, and that S is the disjoint union of non-empty subsets S_0, S_{-1}, S_1 such that $t^{\varepsilon}(S_0 \cup S_{\varepsilon}) \subseteq S_{\varepsilon}$ (for $\varepsilon = \pm 1$), and $gS_{\varepsilon} \subseteq S_0$ for $g \in G \setminus A_{\varepsilon}$, $\varepsilon = \pm 1$. Then G^* is the HNN-extension of G with respect to the isomorphism $A_1 \to A_{-1}$.

3.3. Bipolar Structures. Another important feature uniting free constructions is the existence of a bipolar structure.

Definition. A bipolar structure on a group G is a decomposition of G as a union of disjoint subsets F, EE, EE^*, E^*E, E^*E^* satisfying the following requirements. (When letters X, Y, Z are used for E or E^*, we assume that $(X^*)^* = X$, *etc.*)

1. F is a subgroup of G.
2. If $f \in F$ and $g \in XY$, then $fg \in XY$.
3. If $g \in XY$, then $g^{-1} \in YX$.
4. If $g \in XY$ and $h \in Y^*Z$, then $gh \in XZ$.
5. For every g in G there exists a natural number $N(g)$ such that whenever $g = g_1 \ldots g_n$ for elements $g_1, \ldots, g_n$ of G and $X_0, \ldots, X_n$ are such that $g_i \in X_{i-1}^* X_i$ for $i = 1, 2, \ldots, n$, it follows that $n \leqslant N(g)$.
6. $EE^* \neq \varnothing$.

For a free product $G = G_1 * G_2$ with amalgamated subgroups A_1 and A_2 such that $A_1 \neq G_1$ and $A_2 \neq G_2$, we have the following natural bipolar structure: $F = A_1 = A_2$. Further, ever element of $G \setminus F$ has an expression in G of the form $g = g_1 \ldots g_n$, where $g_1, \ldots, g_n$ lie in $G_1 \setminus F$ or $G_2 \setminus F$. We say that $g \in EE$ if $g_1 \in G_1$ and $g_n \in G_1$; $g \in EE^*$ if $g_1 \in G_1$, $g_n \in G_2$ *etc.* In the case of an HNN-extension G^* of a group G with respect to the isomorphism $\phi \colon A \to B$ of subgroups, for F we take A. For each element g of $G^* \setminus F$, consider its reduced expression (see 2.2):

$$g = g_0 t^{\varepsilon_1} g_1 \ldots t^{\varepsilon_n} g_n.$$

In this case, $g \in EE$ if and only if $g_0 \in G \backslash A$ or $g_0 \in A$ and $\varepsilon_1 = 1$ *and* $g_n \in G_n \backslash A$ or $g_n \in A$ *and* $\varepsilon_n = -1$; $g \in EE^*$ if and only if $g_0 \in G \backslash A$ or $g_0 \in A$ and $\varepsilon_1 = 1$ *and* $g_n \in A$ and $\varepsilon_n = 1$; $g \in E^*E$ if and only if $g_0 \in A$ and $\varepsilon_1 = -1$, $g_n \in G \backslash A$ or $g_n \in A$ and $\varepsilon_n = -1$; $g \in E^*E^*$ if and only if $g_0 \in A$, $\varepsilon_1 = -1$, $g_n \in A$ and $\varepsilon_n = -1$.

A group G can carry different bipolar structures, say when G is obtained as two different free constructions. Thus, the fundamental group of the 2-sphere with two handles,

$$G = \langle a_1, a_2, a_3, a_4; a_1 a_2 a_1^{-1} a_2^{-1} = a_3 a_4 a_3^{-1} a_4^{-1} \rangle$$

is the free product of two free groups $F(a_1, a_2)$ and $F(a_3, a_4)$ with cyclic subgroups $\langle a_1 a_2 a_1^{-1} a_2^{-1} \rangle$ and $\langle a_3 a_4 a_3^{-1} a_4^{-1} \rangle$ amalgamated. It is also an HNN-extension with conjugate cyclic subgroups $\langle a_2 \rangle$ and $\langle a_3 a_4 a_3^{-1} a_4^{-1} a_2 \rangle$, since

$$G = \langle a_1, a_2, a_3, a_4; a_1 a_2 a_1^{-1} = a_3 a_4 a_3^{-1} a_4^{-1} a_2 \rangle.$$

The fundamental theorem here is a result of Stallings.

Theorem. *A group G has a bipolar structure if and only if it is an* HNN-*extension or a nontrivial free product with amalgamation.*

"Non-triviality" means that the amalgamated subgroup is different from each of the constituent groups. A special case is that of ordinary free products. Stallings's theorem underlines once again the kinship of the different free constructions, and it facilitates the proof of some of their main properties by making available the language of bipolar structures.

Stallings was the first to introduce bipolar structures. He used them to show that groups of cohomological dimension 1 are free (see § 2 of Chapter 2) in the finitely generated case. (The general solution was discovered by Swan.) In passing, it was proved that torsion-free almost free (that is, torsion-free free-by-finite) groups are free. Later, Cohen proved the following result:

Theorem. *Every almost free group is an* HNN-*extension of a tree product of finite groups in which the conjugate subgroups are contained in a single vertex subgroup.*

3.4. Groups Acting on Trees

Definition. The *Cayley graph of a group G* with distinguished set $M = \{a_i | i \in I\}$ of elements is the graph Γ whose vertex set is G, and the pair (g, h) is joined by an edge if $h = ga_i$ for some $a_i \in M$. (It is natural to consider Γ as a coloured graph, with (g, h) having the i-th colour).

Usually, M is taken to be a generating set for G, and the graph is then connected. Fig. 2 depicts the graph of the symmetric group S_3 generated by the cycles $a = (1, 2)$ and $b = (1, 2, 3)$; Fig. 3 is a fragment of the graph of the free group $F(a, b)$, which is an infinite tree because the elements of $F(a, b)$ have unique uncancellable expressions in terms of a and b.

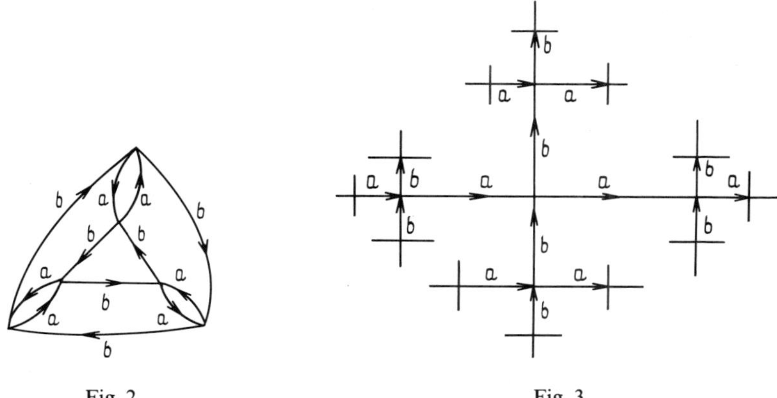

Fig. 2 Fig. 3

Clearly, the left translations $L_g(x) = gx$ for $x \in G$ are automorphisms of Γ, and the map $g \to L_g$ is a homomorphism $G \to \operatorname{Aut} \Gamma$ under which $g \to L_g$ defines a faithful action (that is, only the identity acts trivially) of G on Γ. In particular, a free group F acts on some tree in such a way that every element $g \in F \setminus 1$ fixes no vertex. As Serre has shown, the converse is also true:

Theorem. *If G is a group acting on a tree in such a way that the stabilizer of every vertex is the identity, then G is free.*

Of course, the Nielsen-Schreier theorem on the freeness of subgroups of free groups follows from this.

The principal free constructions were subsumed by Serre in a generalised concept of fundamental group of a *graph of groups*, and it turns out that these groups have natural interpretations as groups acting on trees; from it there follows a theorem about the structure of subgroups of free constructions.

Let Γ be a connected oriented graph containing with every edge e its inverse edge e^{-1}, so that if $\alpha(e)$ and $\omega(e)$ are the start and end of e, then $\omega(e)$ and $\alpha(e)$ are the start and end of e^{-1}. Suppose that to every vertex v in the vertex set Γ^0 there is associated a "vertex group" G_v, to every edge e in the edge set Γ^1 there is associated an "edge group" G_e, and there are monomorphisms $g \to g_+$ and $g \to g_-$ of G_e into $G_{\alpha(e)}$ and $G_{\omega(e)}$ respectively. The system $(\mathfrak{G}, \Gamma)$ of groups and monomorphisms is called a *graph of groups*. Further, we extract a maximal tree T from Γ, whose existence is guaranteed by Zorn's lemma.

Definition. The *fundamental group* $\pi(\mathfrak{G}, \Gamma, T)$ of the *graph* $(\mathfrak{G}, \Gamma)$ *of groups* with respect to the maximal tree T is the group on the generating set M, where M is the disjoint union of generating sets for the vertex groups G_v $(v \in \Gamma^0)$ and a set of symbols t_e for $e \in \Gamma^1 \setminus T^1$, with defining relations of three types:

1) defining relations for the G_v, $v \in \Gamma^0$;
2) relations $g_+ = g_-$ for $g \in G_e$, $e \in T^1$;
3) relations $t_e g_+ t_e^{-1} = g_-$ for $g \in G_e$, $e \in \Gamma^1 \setminus T^1$.

As is clear from the definition, the free product with amalgamation and the HNN-extension are fundamental groups of the graphs of groups depicted in Figs. 4 and 5 respectively. In the case of a tree product $T = \Gamma$, that is, Γ is a tree.

Fig. 4 Fig. 5

The Bass-Serre theorem is fundamental here. It states that every group acting on a tree X without inverse edges (which means that equations of the form $e^{-1} = ge$ are impossible) is the fundamental group of the graph $(\mathfrak{G}, \Gamma, T)$ of groups constructed as follows. T is the lifting to X of a maximal tree in the quotient-graph $Y = G\backslash X$; $\Gamma^0 = T^0$, and Γ^1 is the connected lifting of the edges of Y containing T. The start and end of the edge e in Γ are defined to be the vertices of T G-equivalent to the start and end of the edge e in X. The vertex groups G_v are the stabilisers in G of the vertices $v \in \Gamma^0 = T^0$, the edge groups G_e are the stabilisers in G of the edges $e \in \Gamma^1$. For $e \in T^1$, the monomorphisms from G_e into $G_{\alpha(e)}$ and $G_{\omega(e)}$ are the natural embeddings $G_e \subseteq G_{\alpha(e)}$, $G_e \subseteq G_{\omega(e)}$. The symbols "+" and "−" depend on the choice of starts and ends of edges. If $e \in \Gamma^1 \backslash T^1$, $\alpha(e) \in T^0$, $\omega(e) \notin T^0$, then $\omega(e)$ is equivalent to a unique vertex $w \in T^0$. Let t_e be an element of G such that $t_e\omega(e) = w$. The monomorphism $G_e \xrightarrow{+} G_{\alpha(e)}$ is the usual embedding $G_e \subseteq G_{\alpha(e)}$, and $G_e \to G_{\omega(e)}$ is obtained from conjugating the embedding $G_e \subseteq G_{\omega(e)}$ by t_e. The case $\alpha(e) \notin T^0$, $\omega(e) \in T^0$ is to be considered in the symmetric fashion.

Conversely, the fundamental group G of a graph $(\mathfrak{G}, \Gamma, T)$ of groups acts by left translations on the coset tree X. Its vertices are the left cosets of the form gG_v, $g \in G$, $v \in \Gamma^0$, the edges are the cosets gG_e^+, $e \in \Gamma^1$, $g \in G$. Here $\alpha(gG_e^+) = gG_{\alpha(e)}$ and $\omega(gG_e^+) = gG_{\omega(e)}$ for $e \in T$, and $\omega(gG_e^+) = gt_eG_{\omega(e)}$ for $e \in \Gamma^1 \backslash T$.

Thus we have:

Theorem. *Every subgroup of the fundamental group of a graph* $(\mathfrak{G}, \Gamma, T)$ *of groups is the fundamental group of a graph of groups whose vertex and edge groups are isomorphically embedded in the vertex and edge groups of* $(\mathfrak{G}, \Gamma, T)$.

This assertion embraces, in particular, the theorem of Nielsen and Schreier (1.2), Kurosh, and Karras and Solitar (2.2).

The structure of groups acting on trees has been used by Serre and Bass in their study of GL_2 and SL_2 over local fields.

§4. Finitely Presented Groups

4.1. Classical Algorithmic Problems in Group Theory. Practically every problem that can be posed for a group G given by a presentation $\langle X; R \rangle$ may present serious difficulties, even if G is finitely presented. The reason is that the determination of equality of words in G is nonconstructive. The fundamental algorithic problems for groups were posed by Dehn in 1911. Suppose that we are given a presentation of a group $G = (X, R)$ (of a group H):

1) for an arbitrary word w in the alphabet X, decide in finitely many steps whether w is equal to the identity in G;

2) for any pair v, w of words, decide whether or not they are conjugate in G;

3) for arbitrary H, recognise in a finite number of steps whether or not it is isomorphic to G.

These questions are important for applications. The first two correspond to the algorithmic problems of the contractibility of a loop to a point in a topological space X and the isotopy of two loops in X. The third question is connected with the homeomorphism problem for two spaces. Problem 1) is called the word (identity) problem in G (or the problem of identity of words). 2) is the conjugacy problem, and the third is the isomorphism problem for G.

We shall not be using a formal concept of algorithm in this survey. It is enough for the reader to have an intuitive concept of algorithm and the certainty that the proper definitions exist. It is important to note, however, that the posing of algorithmic problems becomes exact only when the concept of algorithm has been formalised (Turing). Algorithmic questions for groups are usually expressed now as problems of the existence of algorithms of this or that sort.

The papers of Novikov from the period 1952–1955 are pioneering works in this area of group theory.

Theorem. *There is a finitely presented group for which there is no algorithm deciding the word problem.*

Indeed, examples can be constructed explicitly. Analogous groups were found by Boone in papers from 1954–57. Novikov's group, and a simpler example found subsequently by Britton, are obtained as free products with amalgamation of groups with soluble word problems. Of course, the conjugacy problem for these groups is *a fortiori* insoluble, since conjugacy to the identity is the same as equality. Novikov proved that the isomorphism problem is also insoluble, that is, it is impossible to decide algorithmically whether arbitrary pairs of finitely presented groups are isomorphic or not. Later, Adian showed that for every fixed group G, there is no algorithm for deciding whether an arbitrary group H is isomorphic to G or not. For example, there is no general algorithm for deciding whether a group is trivial: $G \cong \{1\}$?

Markov established the algorithmic non-recognisability of a wide class of properties of semigroups; they came to be known as Markov properties.

Definition. A *property* of a group is said to be a *Markov* property if 1) it is preserved under isomorphisms (that is, it is an abstract property); 2) there is a group possessing the property; 3) there exists a group not embeddable in any group with the property.

Adian (and later Rabin) obtained a group analogue of a semigroup theorem of Markov.

Theorem. *Every Markov property of groups is algorithmically non-recognisable.*

Examples of Markov properties are finiteness, commutativity, and freeness, as well as many others.

The Higman and Boone-Higman theorems reveal a stable connection between algorithmic problems and general questions about the structure of finitely presented groups. In order to state the first of these, we recall that a subset R of a set S is said to be recursive if there exists an algorithm for recognising whether any element s of S lies in R or not. A set R is recursively enumerable if there is an algorithm enabling the elements of R to be enumerated (by the natural numbers) in some order. A *presentation* is said to be *recursive* if the generating set X is finite and the set R of defining relations is recursively enumerable.

Theorem. *A finitely generated group G is a subgroup of a finitely presented group if and only if G has a recursive presentation.*

From the Boone-Higman result there follows:

Corollary. *The word problem in a finitely generated group G is algorithmically soluble if and only if G is isomorphically embeddable in a simple subgroup of some finitely presented group.*

It is interesting to note that there is a finitely presented group that embeds every recursively presented group.

Nielsen's method (section 1.3) makes it possible to give an algorithm that decides, for each element of a free group F, whether or not it lies in the subgroup of F generated by a given set $\{w_1, \dots, w_n\}$; that is, the algorithmic occurrence problem for F has the positive solution. However, as Mikhailova has shown, the occurrence problem has the negative solution even for the direct product $F \times F$. We note here a result of Moldovanskij.

Theorem. *There exists an algorithm that recognises whether the subgroups $\langle U \rangle$ and $\langle V \rangle$ generated by two finite subsets U and V of a free group F are conjugate or not (and whether $\langle U \rangle$ is conjugate to a subgroup of $\langle V \rangle$).*

4.2. One-Relator Groups. The first important examples of one-relator groups arose as fundamental groups of compact closed 2-manifolds. The relations take the form $\prod_{i=1}^{g} a_i b_i a_i^{-1} b_i^{-1} = 1$ or $\prod_{i=1}^{n} a_i^2 = 1$ according as to whether the surface is orientable or non-orientable.

The foundations for the theory of one-relator groups were laid by Magnus. His method is applicable to the study of the wider class of groups having so-called graded presentations; such a presentation means that the group can be built up from groups with shorter defining relations using free products with amalgamations (or HNN-extensions: this is a later treatment based on a paper by Moldovanskij (1967)).

As an illustration, we consider the group $G = \langle a, b; a^2 b^4 a^{-2} b^2 = 1 \rangle$. It can be mapped homomorphically onto an infinite cyclic group $\langle a \rangle$ *via* the map $a \mapsto a$, $b \mapsto 1$. The Reidemeister-Schreier rewriting process (section 2.1) gives generators $b_k = a^k b a^{-k}$ for the kernel N of this homomorphism, and a presentation

$$N = \langle \dots b_{-2}, b_{-1}, b_0, b_1, b_2, \dots; b_i^4 b_{i-2}^2 = 1, i = 0, \pm 1, \dots \rangle.$$

The defining relations $r_i = 1$ of N are shorter than those defining G. They form a graded system in the sense that, for $i < j$, the smallest numbers k_i and k_j and the largest numbers l_i and l_j of generators occurring in the cyclically uncancellable forms of r_i and r_j satisfy the inequalities $k_i < k_j$ and $l_i < l_j$. The group N is made up from the groups

$$N_i = \langle b_{i-2}, b_{i-1}, b_i; b_i^4 b_{i-2}^2 = 1 \rangle$$

via the construction of a tree product with the free subgroups $\langle b_{i-1}, b_i \rangle$ of N_i and N_{i+1} amalgamated, for $i = 0, \pm 1, \dots$. In inductive arguments elucidating properties of G, it can be assumed that the groups N_i have the relevant properties.

If $r \equiv uv$, then vu is said to be a cyclic rearrangement of the word r. A word r is cyclically reduced if all its cyclic rearrangements are uncancellable. It is clear that every group relation is equivalent to a relation $r = 1$, where r is cyclically reduced. We are now in a position to state Magnus's theorem.

The Freiheitssatz. *If $G = \langle a_1, \dots, a_n; r = 1 \rangle$ is a group with a single defining relation, where r is cyclically reduced and contains the letters a_n or a_n^{-1}, then the elements $a_1, \dots, a_{n-1}$ form the basis of a free subgroup of G.*

Another theorem of Magnus provides an algorithm for reconising the identity of words in a one-relator group. (We note that the algorithmic conjugacy and isomorphism problems for arbitrary one-relator groups are still open.) It has also been shown that the normal closure $\langle r_1 \rangle^F$ of an element r_1 in a free group F is the same as $\langle r_2 \rangle^F$ if and only if r_1 is conjugate in F to r_2 or r_2^{-1}. If $G = \langle X; r \rangle$, where $r = s^m$ in $F(X)$, $m \geqslant 1$ and s is not a proper power in F, then s has order m in G, and every element of finite order in G is conjugate in G to some power of s. In particular, G is torsionfree if r is not a proper power in $F(X)$.

In 1969, Moldavanskij described the subgroups of a one-relator group that satisfy a non-trivial identity, and in 1971, Karras and Solitar, and independently Chebotar', described those not containing a subgroup isomorphic with F_2. We note here the interesting example due to Baumslag and Solitar of a non-Hopfian (and therefore not residually finite – see 1.4) one relator group: $\langle a, b; ab^2a^{-1} = b^3 \rangle$.

Notwithstanding the corpus of method for studying one-relator groups (there is nothing similar for more than one relator), there has been no diminution in interest in these groups. To substantiate this claim, we refer to the extraordinarily intriguing problem of Stallings and Jaco about the fundamental group of a sphere with g handles, that is,

$$\Phi_g = \langle a_1, \dots, a_{2g}; \prod_{i=1}^{g} a_{2i-1} a_{2i} a_{2i-1}^{-1} a_{2i}^{-1} = 1 \rangle,$$

the answer to which is equivalent to a proof of the celebrated Poincaré conjecture, which is of the utmost importance for the classification of three-dimensional manifolds. The – as yet unproved – proposition that every compact connected simply-connected three-dimensional manifold is homeomorphic to the three – sphere is equivalent to the assertion that, for every $g \geqslant 2$, every epimorphism $\Phi_g \to F_g \times F_g$ factors through an epimorphism from Φ_g to some free product $A * B$:

$$\Phi_g \to A * B \to F_g \times F_g,$$

where $A \neq \{1\}$, $B \neq \{1\}$.

4.3. Some Classes of Finitely Presented Groups. Geometry and topology are constantly enriching combinatorial group theory with problems. Here we shall simply recall some important sources of examples of finite presentations, after mentioning the origins of the various theorems. The scope of this is so wide, and the character of the material so geometrical, that we shall find it necessary to explain the items individually.

The most thoroughly worked area in the general problem of classifying embeddings of a topological space X in a space Y is that of knot theory. A knot is a subset of Euclidean 3-space $\mathbb{R}^3$ homeomorphic to a circle. Knots K_1 and K_2 are equivalent if one can be carried to the other by a homeomorphism from $\mathbb{R}^3$ to itself. The principal results refer to "tame" knots (Crowell and Fox (1963)), that is, knots equivalent to polygonal knots, knots consisting of finitely many straight-line segments.

The *group* of the knot K is the fundamental group of $\mathbb{R}^3 \setminus K$. (The group of links is defined in a similar way). A necessary condition for the equivalence of knots is that their groups be isomorphic. The group of the trivial knot (a simple circle) is infinite cyclic; the group of the "clover-leaf" (Fig. 6) has two simple presentations:

$$\langle x, y; xyx = yxy \rangle \cong \langle a, b; a^2 = b^3 \rangle.$$

It follows from this that the clover-leaf cannot be untied, that is, is not equivalent to the trivial knot. See Crowell and Fox (1963).

Another frontier region between algebraic topology and combinatorial group theory is the study of braids, begun by Artin in 1925. A good presentation of the theory of braids is to be found in the survey of Magnus, Karras and Solitar (1966). Here, we shall just give defining relations for the *Artin braid group*

Fig. 6

on n threads:

$$B_n = \langle a_1, \ldots, a_{n-1};\, a_{i+1}a_ia_{i+1} = a_ia_{i+1}a_i,\ i = 1, \ldots, n-2,\ a_ia_j = a_ja_i \text{ for } |i-j| > 1\rangle.$$

This group B_n has a faithful representation in the group $\operatorname{Aut} F_n$ of automorphisms of the free group F_n, in which a_i corresponds to the automorphism

$$x_i \mapsto x_{i+1}, \quad x_{i+1} \mapsto x_{i+1}^{-1}x_ix_{i+1},\ x_j \mapsto x_j \quad \text{for} \quad j \neq i, i+1,$$

and this gives an automatic solution of the word problem in B_n. The conjugacy problem in B_n was solved by Garside in 1969.

The braid group B_n is related to the mapping-class groups of the 2-sphere with n points removed. The mapping-class group $M(n, g)$ of a two-dimensional orientable surface R_g of genus g with n points removed is defined to be the factor-group of the group of autohomeomorphisms of R_g that preserve orientation and a fixed set of n points by the path-connected component of the identity in this group. See Magnus, Karras and Solitar (1966) for a representation of $M(n, 0)$ as a factor-group of B_n.

The groups $M(n, g)$ have good embeddings into the outer automorphism groups of Fuchsian groups. For the definition and a description of these, see Lyndon and Schupp (1977). The geometrical method for finding defining relations for discrete groups of transformations using decompositions of spaces into fundamental regions (see Lyndon and Schupp (1977)) also lie outside the scope of this survey.

We conclude the section by returning to knots, and remark that the group G of a tame knot can be given by a presentation with one fewer relations than generators. It follows that the factor-group of G by its commutator subgroup $G' = [G, G]$ is infinite. In fact, G/G' is always cyclic, so that $G'/(G')'$ can be viewed as a module over the ring of Laurent polynomials with integer coefficients. The most important invariants of these modules are the Alexander matrices and polynomials, which are a powerful tool for distinguishing knots (see also section 3 of Chapter 2). The Fox free differential calculus is a method for computing Alexander matrices. Differentiation in the group ring of a free group is applied in other situations to a study of "relation modules" of presentations. Since the Fox derivatives are useful also in a study of soluble groups and identities in groups, we introduce the requisite concepts at this point.

Definition. The *group algebra* $A[G]$ of a group G over a commutative ring A with identity is the associative A-algebra in which the elements of G form a basis over A, and multiplication of basis elements is just as in G.

The homomorphism $\varepsilon\colon A[G] \to A$ taking all elements in the basis G to 1, so that $\varepsilon(\sum_g \alpha_g g) = \sum_g \alpha_g$, is called the *augmentation homomorphism*. Its kernel is the *augmentation ideal* and has basis $\{1 - g\}_{g \in G\setminus\{1\}}$. Any A-linear map $D\colon A[G] \to A[G]$ such that $D(uv) = Du \cdot \varepsilon(v) + uDv$ is called a *differentiation* of the group algebra.

Let $G = F(X)$ be a free group, where $X = \{x_i | i \in I\}$. Then every map $X \to A[G]$ determines exactly one differentiation. The *Fox partial derivatives* $\dfrac{\partial u}{\partial x_i}$ for $x_i \in X$ are defined by the formulae $\dfrac{\partial x_i}{\partial x_i} = 1$ and $\dfrac{\partial x_j}{\partial x_i} = 0$ for $j \neq i$. For every differentiation D, we have $Du = \sum_i \dfrac{\partial u}{\partial x_i} Dx_i$.

We have to make a remark of a general nature at this point. If A is an abelian normal subgroup of a group G, the value of gag^{-1} for an element a of A depends only on the coset gA, so that the factor-group $B = G/A$ acts by conjugation on A. This action can be extended to the integral group ring $\mathbb{Z}[B]$; namely, if $a \in A$ and $z = \sum_{b \in B} \alpha_b b$, then

$$z \circ a = \prod_{b \in B} g a^{\alpha_b} g^{-1} \qquad (b = gA). \tag{1}$$

In this situation, it is customary to write A additively, and formula (1) is easily shown to turn A into a (left) $\mathbb{Z}[B]$-module (the axioms for modules are just like those for vector spaces; the only difference is that the operator domain need not be a field).

In particular, the *relation module* for G with presentation $G = F/N$, F free, is the G-module $\overline{N} = N/N'$, where $N' = [N, N]$ is the commutator subgroup of N. The terminology derives from the fact that when N is the normal closure of a system of defining relations r_i, $i \in I$, $\overline{N}$ is generated as G-module by the cosets $r_i N'$.

The isomorphic representation of F/N' discovered by Magnus can be defined as follows: take $F = F(X)$. Let M be the free G-module with basis $\{b_i\}_{i \in I}$ (that is, every element of M has a unique expression in terms of the basis elements with coefficients in $\mathbb{Z}[G]$), where $G = F/N$. We consider the group of matrices of the form $\begin{pmatrix} g & u \\ 0 & 1 \end{pmatrix}$, where $g \in G$, $u \in M$, and multiplication is according to the rule

$$\begin{pmatrix} g_1 & u_1 \\ 0 & 1 \end{pmatrix} \begin{pmatrix} g_2 & u_2 \\ 0 & 1 \end{pmatrix} = \begin{pmatrix} g_1 g_2 & g_1 \circ u_2 + u_1 \\ 0 & 1 \end{pmatrix},$$

where "$\circ$" denotes module multiplication. We denote by $\bar{a}$ the natural image of the element a of F/N' in F/N; the extension of this map to an epimorphism $\mathbb{Z}[F/N'] \to \mathbb{Z}[G]$ of algebras is also denoted by the bar. The *Magnus embedding* is given by the formula

$$\mu(g) = \begin{bmatrix} \bar{g} & \sum_i \dfrac{\partial g}{\partial x_i} \circ b_i \\ 0 & 1 \end{bmatrix}.$$

where $g \in F/N'$. As Remeslennikov and Sokolov have remarked, a necessary and sufficient condition for the matrix $\begin{pmatrix} g & \sum u_i \circ b_i \\ 0 & 1 \end{pmatrix}$ to lie in the image $\mu(F/N')$ is that $g - 1 = \sum_i u_i(\bar{x}_i - 1)$. The Magnus embedding is widely used to study groups of the form F/N'; in particular, free soluble groups (see Chapter 2).

§ 5. Cancellation Theory

5.1. Groups with Small Cancellation Conditions. When he solved the word problem for the fundamental group Φ_g of a two-dimensional closed orientable surface of genus g, $g \geqslant 2$, Dehn noticed that the graph of Φ_g is obtained by covering the Euclidean plane with $4g$-gons in such a way that the degree of every vertex is $5g$. If an arbitrary finite number of polygons is chosen for such a covering, an easy application of Euler's formula enables one to find a polygon with more than half of its edges not lying in any other polygon in the chosen subset; that is, they lie on the boundary of the figure under consideration. It follows from this that a nonempty uncancellable word w equal to the identity in Φ_g must contain a subword u which is simultaneously a subword of a cyclic rearrangement of r or r^{-1}, where $r \equiv \prod_{i=1}^{g} a_{2i-1}a_{2i}a_{2i-1}^{-1}a_{2i}^{-1}$ is one of the defining relations for Φ_g such that $|u| > \frac{1}{2}|r|$. Since $r = 1$, it is clear that $u = v$ in Φ_g, where $|u| + |v| = |r|$, so that $|v| < |u|$. Thus, an algorithm for deciding the equation $w = 1$ in Φ_g reduces to a process which decreases the length of w monotonically. The *Dehn algorithm* was later extended to a wide class of groups with so-called small cancellation conditions. A paper of Tartakovskii from 1949 is important for a combinatorial-algebraic approach to studying groups with small cancellation conditions; the motivation was an investigation of relations in periodic groups.

Definition. The expression $G = \langle X; R \rangle$ for a presentation is said to be *symmetrised* if the words in R are uncancellable, and each word r in R occurs together with its inverse and every cyclic rearrangement of r.

It is clear that every presentation can be symmetrised.

The "*small cancellation condition*" $C'(\lambda)$ involving a parameter λ, $0 < \lambda < 1$, is one of the easier ones to state.

Definition. A symmetrised presentation $G = \langle X; R \rangle$ satisfies the condition $C'(\lambda)$ if, for every pair of defining words r_i and r_j that are not mutually inverse, complete cancellation of the product $r_i r_j$ deletes less than $\lambda \min(|r_i|, |r_j|)$ letters from each of the factors r_i, r_j.

For example, it is clear that the presentation of Φ_g mentioned in the preceding section satisfies $C'(\lambda)$ for every $\lambda > 1/4g$.

Tartakovskii solved the word problem for $C'(1/6)$-groups by purely algebraic methods, while Grindlinger showed that the Dehn algorithm is applicable to them. In fact, he obtained more precise information on the relations (Lyndon and Schupp (1977)), and also solved the conjugacy problem for $C'(1/8)$-groups.

For small values of λ, a noncyclic group G in $C'(\lambda)$ is very close to being free. It contains many free subgroups and has a very rich lattice of normal subgroups. In such a group, elements a and b commute if and only if they are contained in a single cyclic subgroup. G is torsionfree if the defining words are not proper powers in the free group. Otherwise, every word of finite order is

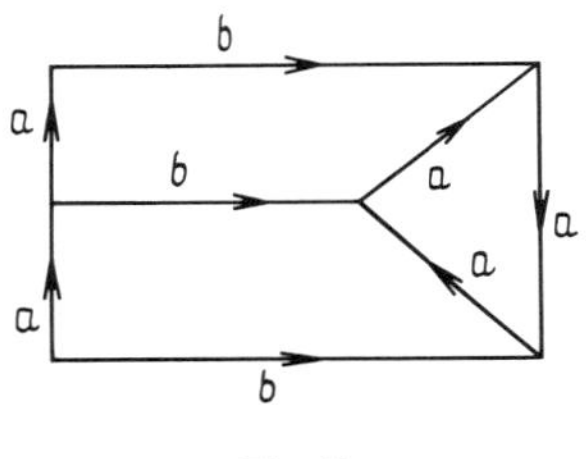

Fig. 7

conjugate in G to a power of some word s such that $s^m \in R$. Extraction of roots in the group is also unique: if an element has an n-th root, it has only one.

See Lyndon and Schupp (1977) in connection with other small cancellation conditions.

5.2. A Geometrical Interpretation of the Deduction of Consequences of Defining Relations. We begin with a figure depicting the deduction of the relation $a^2bab^{-1} = 1$ from the relations $a^3 = 1$ and $aba^{-1}b^{-1} = 1$ (Fig. 7). The interpretation is as follows. On traversing the contour of a quadrilateral or a triangular cell, we read one of the defining words (if an edge goes in the direction opposite to its arrow, we assign it the inverse of the appropriate letter), and on traversing the contour of the whole diagram we read the consequence: a^2bab^{-1}.

In 1933, van Kampen noticed that the analogous interpretation of the deduction of consequences in groups is universal – it is of no importance whether the group is Fuchsian, that is, whether its graph is planar in the large. Indeed, consider the Euclidean plane $\mathbb{R}^2$. A map $\mathcal{M}$ consists of a disjoint union of finitely many subsets of $\mathbb{R}^2$ of three types: vertices, edges and cells. A vertex is a point, an edge is a bounded subset of $\mathbb{R}^2$ homeomorphic to the interval $(0, 1)$ in $\mathbb{R}$, and a cell is a bounded subset of $\mathbb{R}^2$ homeomorphic to the open unit disc. In addition, it is required that the closure of an edge e is $\bar{e} = \{e\} \cup \{u\} \cup \{v\}$, where u and v are vertices, and the boundary of a cell Π is $\partial\Pi = \bar{e}_1 \cup \cdots \cup \bar{e}_n$ for some edges $e_1, \ldots, e_n$.

Suppose now that $G = \langle X; R \rangle$ is a symmetrized presentation.

Definition. A *diagram* over G is any connected simply-connected map $\mathcal{M}$ such that

1) every geometric edge e is interpreted to be a pair of edges e and e^{-1} with opposite orientations;

2) with every edge e there is associated a letter in $X^{\pm 1}$, the label $\phi(e)$ of the edge, satisfying the equation $\phi(e^{-1}) = \phi(e)^{-1}$;

3) if $e_1, \ldots, e_n$ is a closed bounding path of a cell Π (the contour of Π), then $\phi(e_1)\phi(e_2)\ldots\phi(e_n) \in R$.

The contour of the improper cell $\mathbb{R}^2 \setminus M$ is also called a contour of the diagram Δ. van Kampen's observation goes like this.

Lemma. *The relation $w = 1$ holds in a group G if and only if there is a diagram Δ over the presentation $G = \langle X; R \rangle$ such that the label $\phi(e_1)\dots\phi(e_n)$ of the contour $e_1, \dots, e_n$ is letter-for-letter identical with w.*

We can restrict ourselves to reduced diagrams, that is, diagrams in which there is no pair of cells Π_1, Π_2 with a common edge e such that the labels of the contours $ee_2\dots e_n$ and $ee'_2\dots e'_n$ of these cells coincide (see Fig. 8). Such cells can be "reduced" without changing the label of the contour of Δ if they are excised from Δ and the hole thus formed is glued up by identifying e'_2 with $e_2, \dots, e'_n$ with e_n.

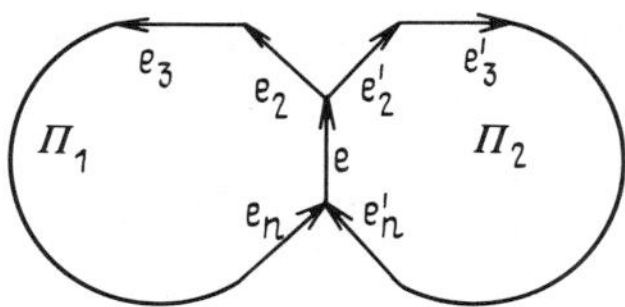

Fig. 8

Unlike the other well-known connections between groups and 2-complexes (but not surfaces!), where every group has a realization as the fundamental group of some space X (see Lyndon and Schupp (1977)), van Kampen's Lemma above went unnoticed until the sixties; at that point, Lyndon and Weinbaum discovered a use for diagrams in small cancellation theory. The fact is that for $G = \langle X; R \rangle \in C'(\lambda)$ (see 5.1), the length k of the common arc $e_1 \dots e_k$ of two cells Π_1 and Π_2 in every reduced diagram over G is less than $\lambda \cdot \min(|\partial\Pi_1|, |\partial\Pi_2|)$, where $|\partial\Pi_i|$ is the length of the contour of Π_i. Thus, for small λ (say, for $\lambda \leqslant 1/6$), every internal cell Π of a diagram Δ has many neighbours, and an easy application of Euler's formula leads to the existence of a cell Π_0 in Δ whose contour has a subpath $e_1 \dots e_m$ occurring in the outer contour of Δ, with $m > \frac{1}{2}|\partial\Pi_0|$. van Kampen's theorem then applies to show that Dehn's algorithm (see 5.1) can be used to solve the word problem in groups satisfying $C'(1/6)$.

Lyndon discovered a natural explanation, as well as other small cancellation conditions generalizing the properties of regular tessellations of the plane. A more detailed study of diagrams led to an improvement of the bound and a discovery of an algorithmic solution to the word problem in $C'(1/5)$. (Dehn's algorithm fails already at this point.) The definitive nature of this results is highlighted by a theorem of Goldberg; he proved in 1978 that every finitely presented group has a finite presentation with $C'(\lambda)$, for each $\lambda > 1/5$. Thus, by Novikov's theorem (see 4.1), each class $C'(\lambda)$ with $\lambda > 1/5$ contains a group with insoluble word problem.

Schupp solved the conjugacy problem in $C'(1/5)$ using ring diagrams (here the simple-connectednedd condition is replaced by the requirement that $\mathbb{R}^2 \setminus \mathcal{M}$ has two components) interpreting the conjugacy relation in the group. In 1972,

Appel and Schupp applied small cancellation theory to solve the conjugacy problem for alternating knot groups. The geometrical methods of small cancellation theory can also be developed over free constructions, that is, further relations are imposed on a free product with amalgamation or HNN-extension. The reader will find proofs of embedding theorems using this method, together with other results, in Lyndon and Schupp (1977).

5.3. Results of Novikov and Adian on Periodic Groups. If a group given by defining relations has no natural representation as a group of automorphisms of some mathematical object, considerations of the interdependence between the defining relations inevitably appear. An intense example of a problem of this sort is the investigation of the *free Burnside group* $\mathbb{B}(m, n)$, that is, the factor-group of the free group F_m by the least normal subgroup containing the n-th powers of all words.

Novikov and Adian answered the *Burnside problem*[2] by establishing the following result.

Theorem. *For* $m \geqslant 2$ *and sufficiently large odd* n (*for example*, $n \geqslant 665$[1]), $\mathbb{B}(m, n)$ *is infinite.*

The theory of transformations of words created by these authors is based on complicated inductive definitions and the proof of a large number of interconnected simultaneous inductive assertions. The difficulty arises because the identity $x^n = 1$ is equivalent to an infinite set of defining relations in $\mathbb{B}(m, n)$. For even the most reasonable choice of relations of this sort, there is inevitably a significant amount of cancellation between them. For instance, if $k \approx n/2$, cancellation of the product of x_1^n and $(x_1^{-k}x_2)^n$ removes about half of the first word.

Obstacles of this sort are not encountered in the proof that 2-generator semigroups with the identity $x^n = 0$ can be infinite – a result that follows immediately from Thue's combinatorial lemma on the existence of arbitrarily long words in the alphabet $\{x, y\}$ not having nonempty subwords of the form w^3. Novikov and Adian also found these considerations useful. However, their assault on the Burnside problem was crowned with success only after they had obtained a classification of periodic words in terms of rank.

Definition. A word w is said to be a *periodic word* with period v if v is a nonempty cyclically reduced word that is not a proper power in the free group, and w is a subword of some power v^l for $l > 0$.

The most important property of periodic words is the uniqueness of the period: if w is a periodic word with periods v_1 and v_2 for which $|w| \geqslant |v_1| + |v_2|$, then v_2 is a cyclic rearrangement of v_1. There follows a small cancellation condition for defining words whose lengths differ "by not too much".

---

[2] For more on the Burnside problem see § 5 of Chapter 2, and in greater detail the survey "Identities" in this series.

Although the groups $\mathbb{B}(m, n)$ with the stated conditions are not finitely presented, there are algorithms solving the word and conjugacy problems for them. Every finite or abelian subgroup of $\mathbb{B}(m, n)$ turns out to be cyclic. The groups contain subgroups that are free Burnside groups of infinite rank (Shirvanyan), which likens them to (absolutely) free groups.

A modification of the relations for $\mathbb{B}(m, n)$ led Adian to a torsionfree group $\mathbb{A}(m, n)$ with infinite cyclic centre and central factor-group isomorphic to $\mathbb{B}(m, n)$. This non-commutative group shares with the additive group $\mathbb{Z}$ of integers the property that every pair of nontrivial subgroups intersect nontrivially. Using an additional factorization, one can get from $\mathbb{A}(m, n)$ a countable non-topologizable group. In conjunction with free products, the method of Novikov and Adian gives new examples of operations on the class of groups without involutions (that is, with no elements of order 2) and new simple groups.

Another applicaton: the beautiful infinite system of identities $[x^{np}, y^{np}]^n = 1$ (p runs over the prime numbers), in which none of the relations follows from the others (Adian (1975)). The technical complexity of the machinery is something of a deterrent to using it. As yet, there has been no real progress on the Burnside problem for small exponents n, nor for 2-powers.

5.4. Topological Approaches to the Construction of Groups with Given Properties. Many conditions on a group can be guaranteed by imposing some relations or other on its elements. For example: periodicity, completeness (or divisibility – see the definition in section 1 of Chapter 2), or the commutativity of all proper subgroups can be achieved *via* specially chosen relations. However, the problem becomes serious when the time comes to prove that the examples so constructed are nontrivial, that is, are not the identity group; or that they are infinite, noncommutative or whatever. The remarks at the end of section 5.1 suggest that the traditional methods of small cancellation theory will not provide ways of establishing properties of this sort.

Nevertheless, geometrical interpretation of the deduction of consequences to defining relations (see 5.2) turns out to be very fruitful here as well, and it helps to clarify the essence of the situation to the same extent as Lyndon's approach explained the theorems of Tartakovskii and Grindlinger. It seem that this effect can be attributed to the fact that diagrams contain all the essential information about the deduction of consequences, while the visually obvious Jordan curve theorem gives poor algebraic information.

Ol'shanskij has made essential use of elementary topological considerations in the solution of purely algebraic questions. To those ends he has constructed presentations of groups with absolutely new properties, for example:

Theorem. *There exist (effectively constructible) infinite nonabelian groups whose proper subgroups are all infinite cyclic.*

In connection with the Shmidt problem about the existence of infinite nonabelian groups with all proper subgroups finite, we make the following definition.

Definition. A group G is said to be *quasifinite* if it is infinite and all its proper subgroups are finite.

It is easy to check that the only abelian quasifinite groups are the quasicyclic groups C_{p^∞} (see section 1 of Chapter 2).

Theorem. *There exist (continuously many) nonabelian quasifinite groups.*

There are effective constructions for such groups, and their algorithmic word and conjugacy problems have positive solutions. Deryabina and Ol'shanskij have shown that a finite group G can be embedded in a quasifinite group if and only if G is a direct product of a group of odd order and an abelian subgroup of order 2^k. We note the special position of the number two here, and recall that a p-group is a group whose elements all have orders that are powers of the prime p. As Shmidt showed, there are no quasifinite 2-groups; however, examples exist for $p \geqslant 3$ (Deryabina).

Ol'shanskij has found a proof of the Novikov-Adian theorem that is significantly shorter than theirs, though with a worse lower bound for the exponent n. He has also constructed an infinite group G with two strong finiteness conditions: for every sufficiently large prime p, there exists a group G in which every element satisfies the equation $x^p = 1$, and every proper subgroup is cyclic. This solves a problem of Tarski. With these results as background, Atabekyan has recently strengthened the theorem about subgroups of $\mathbb{B}(m, n)$ mentioned in 5.3: for sufficiently large odd n, every noncyclic subgroup contains a subgroup isomorphic to $\mathbb{B}(m, n)$.

Guba has drawn on some additional considerations and constructed the first example of a finitely generated divisible group (a similar result has been announced by Rips). He has also recently found some finitely generated simple groups that are not 2-generator.

The presentations $G = F/N$ for all the groups cited above are aspherical, that is, there do not exist reduced spherical van Kampen diagrams over them (reduced in a certain generalised sense, that is) (see 5.2). This fact enables one to give explicit descriptions of the universal central extensions $F/[F, N]$ and the universal abelian extensions $F/[N, N]$ for groups $G = F/N$ of the sort under discussion. Using this, Zyabrev and Reznichenko have constructed a path-connected topological group in which a neighbourhood of the identity satisfies a nontrivial identical relation that cannot be extended to the whole group (this was a question of Platonov). An identity of the form $x^n = 1$ turned out to be suitable.

These new presentations are constructed inductively, and their properties are verified by induction, just as in the work of Novikov and Adian (see 5.3). The difference is that the typical objects arising are diagrams (of the usual type, and also diagrams with some "holes", diagrams on tori and the like), paths in them that are close to being geodesics, configurations of cells, *etc*. The following visually intuitive geometrical fact is important here. Suppose that a sphere is divided into cells such that the length of the arc p common to two cells Π_1 and

Π_2 never exceeds the semiperimeters $|\partial\Pi_i|$, $i = 1, 2$, and, if $|\partial\Pi_1|$ and $|\partial\Pi_2|$ are comparable in size (in a certain exact sense: for example, $|\partial\Pi_1| < 1000|\partial\Pi_2|$ and $|\partial\Pi_2| < 1000|\partial\Pi_1|$), then $|p| \ll |\partial\Pi_i|$, $i = 1, 2$. Then there are cells Π and Π' between which there are "squeezed" many cells with perimeters that are small in comparison with $|\partial\Pi|$ and $|\partial\Pi'|$ (see Fig. 9).

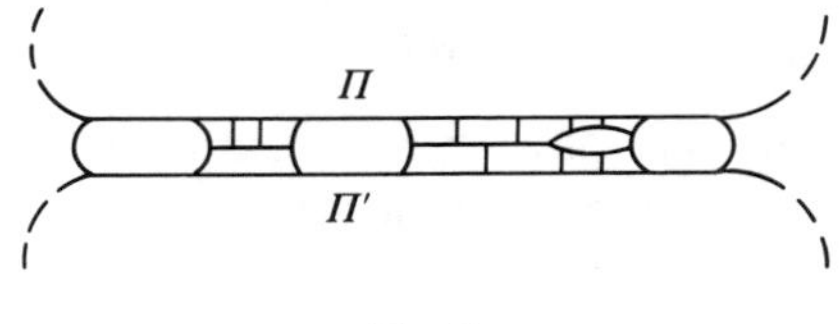

Fig. 9

§ 6. Other Combinatorial Problems

6.1. Equations over Groups

Definition. By an *equation* in n variables with *coefficients in a group G* we mean an equation

$$w(x_1, \ldots, x_n; g_1, \ldots, g_m) = 1 \tag{1}$$

where w is an element of the free product $G * F(x_1, \ldots, x_n)$ and $g_1, \ldots, g_m \in G$.

If G is a subgroup of a group K, a *solution* of equation (1) is any n-tuple $(a_1, \ldots, a_n)$ of elements of K such that $w(a_1, \ldots, a_n; g_1, \ldots, g_m) = 1$ in K; that is, w goes to the identity under the homomorphism $G * F(x_1, \ldots, x_n) \to K$ carrying each x_i to a_i and acting as the identity on G. If $K = G$, we speak of solutions to (1) in G.

The root-adjunction problem is that of embedding G in some group K in which (1) has a solution. For example, roots of the equation $x^k = g$ can always be adjoined to G in a free product with cyclic amalgamation: $K = \langle G, \langle x\rangle = \langle x^k\rangle\rangle$. This process, coupled with transfinite induction, enables one to show that every group G can be embedded in a complete (= divisible) group $\bar{G}$, that is, one where every element has a k-th root for every k.

It is easy to see that adjunction of roots is not always possible. For example, the equation $xg_1x^{-1} = g_2$ is insoluble in every supergroup of G if g_1 and g_2 have different orders. If g_1 and g_2 have the same order, there is a solution; for example, in an HNN-extension of G with respect to an isomorphism of the cyclic subgroups $\langle g_1\rangle$ and $\langle g_2\rangle$.

Gerstenhaber and Rothaus have obtained the following analogue of Cramer's theorem:

Theorem. *Let G be a subgroup of a compact connected Lie group. Then a system $w_1 = 1, \ldots, w_n = 1$ of n equations in n unknowns $x_1, \ldots, x_n$ has a solution over G if the determinant of the matrix composed of the sums a_{ij} of the exponents of the x_i in w_j is nonzero.*

The widest class of groups to date over which every equation has a solution was discovered by Brodsky in 1980. These are the groups such that all nontrivial finitely generated subgroups have homomorphisms onto the infinite cyclic group. Free groups, torsionfree one-relator groups, and many classes of soluble groups, linear groups *etc* are of this type.

Definition. A group G is said to be *algebraically closed* if every finite set of equations that are simultaneously soluble over G has a solution in G itself.

Every group is a subgroup of an algebraically closed group. Algebraically closed groups are simple and are not finitely generated: this was proved by B.H. Neumann. He also established the following result.

Theorem. *Every finitely generated group with soluble word problem can be embedded in every algebraically closed group.*

McIntyre proved the converse:

Theorem. *If a finitely generated group G can be embedded in every algebraically closed group, the word problem is soluble in G.*

Thus, one can give a criterion in group-theoretical terms for the algorithmic solubility of the word problem. Miller has shown that algebraically closed groups are non-constructive, that is, no algebraically closed group has a recursive (infinite) presentaton.

In 1980, Brodsky established the existence of a non-trivial group such that every equation soluble over the group has a solution in it.

6.2. Equations in a Free Group. A description of solutions of equations in a free group $F(X)$ has been achieved only in very special cases. If $a, b \in F(X)$, $a \neq b$, and (u, v) is a solution of the equation

$$xyx^{-1}y^{-1} = aba^{-1}b^{-1},$$

then $(uv^{\pm 1}, v)$ and $(u, vu^{\pm 1})$ are also solutions. In 1962, Mal'tsev showed that every solution of this equation in $F(X)$ can be obtained by finitely many appropriate elementary transformations on the solution (a, b). Wicks has observed that the equation $xyx^{-1}y^{-1} = g$ has a solution in $F(X)$ for an uncancellable word g if and only if g is conjugate to a cyclically reduced word letter-for-letter equal to a product of the form $uvwu^{-1}v^{-1}w^{-1}$.

The definitive result for equations in a single variable is due to Lorenz.

Theorem. *The set of solutions of a finite system of equations in a single variable in a free group is given by a finite set of parametric words of the form uv^kw, $k = 0, \pm 1, \dots$.*

The analogous assertion for several unknowns is false.

There is a theorem of Baumslag and Steinberg relating to equations of the form

$$w(x_1, \dots, x_n) = g^m, \qquad m > 1, \tag{2}$$

where w is not part of any basis of $F(x_1, \ldots, x_n)$ and is not a proper power in that group.

Theorem. *If* $(g_1, \ldots, g_n)$ *is a solution of equation* (2) *in the free group* F, *then the rank of* $\langle g_1, \ldots, g_n \rangle$ *is at most* $n - 1$.

In particular, elements u, v, w of a free group permute in pairs if they are connected by an equation like $u^k v^l = w^m$, where $k, l, m \geqslant 2$.

Lyndon obtained several results about solutions of equations in a free group by considering certain soluble factor-groups of it. For example, for elements $u_1, \ldots, u_n$ of a free group connected by the relation $u_1^k \ldots u_n^k = 1$, where $k > 1$, the free rank of $\langle u_1, \ldots, u_n \rangle$ is not more than $n/2$. Edmunds and Piollet have studied quadratic equations, that is, equations containing two occurrences of every variable.

Makanin proved the following theorem in answer to a well-known question of Tarski.

Theorem. *There exists an algorithm that decides of an arbitrary equation whether or not it has a solution in a free group.*

In the proof, an explicit construction is given of a function $\Phi(\delta)$ of a natural argument such that, if an equation is soluble in a free group $F(X)$ and $|w| < \delta$, then there is a solution $(a_1, \ldots, a_n)$ with $|a_i| < \Phi(\delta)$ for $i = 1, \ldots, n$. In view of this restriction, the verification that an equation has a solution reduces to an easy enumeration of a finite set of integers. Significant combinatorial difficulty has to be overcome in the construction of $\Phi(\delta)$. We mention just one part of it. The problem brings in a free semigroup on the "double" alphabet $x_1, \ldots, x_\omega$, $x_1^{-1}, \ldots, x_\omega^{-1}$. Equations in it are practically indistinguishable from equations in a free group; however, on substitution of the solution, the left and right hand sides are letter-for-letter identical (not just to within cancellation). The "method of generalized equations" is used; it was created earlier by the author for investigating equations in semigroups.

More general is the unsettled problem of the algorithmic solubility of the elementary theory of a noncyclic free group. It is also unknown whether these theories are the same for free groups of different ranks, that is, whether they can be distinguished in the language of first-order predicate calculus.

In conclusion, we record a somewhat unexpected version of the Hilbert Basis Theorem, discovered recently by Guba.

Theorem. *Every system of equations in n variables in a free group is equivalent to a finite system.*

The proof amounts to a simple but elegant reduction of this assertion to a modern version of Hilbert's theorem using Sanov's linear representation (see 1.4).

6.3. Growth Functions of Groups. Suppose that a generating set $M = \{a_1, \ldots, a_m\}$ is distinguished in a group G.

Definition. The growth function of G relative to M is the function $\gamma(n)$ of a natural argument counting the number of elements of G that can be expressed as products of not more than n factors $a_i^{\pm 1}$, $i = 1, \ldots, m$.

For instance, the direct product $\langle a_1 \rangle \times \langle a_2 \rangle$ of two infinite cyclic groups has growth function $\gamma(n) = 2n^2 + 2n + 1$; for the free group $F(a_1, \ldots, a_m)$,

$$\gamma(n) = \frac{m}{m-1}((2m-1)^n - 1) + 1.$$

The groups $\mathbb{B}(m, n)$ considered in 5.3 also have exponential growth functions (see Adian (1975)).

Functions γ and δ are regarded as equivalent if there is a constant c such that, for all n,

$$\gamma(n) \leqslant c\delta(cn), \qquad \delta(n) \leqslant c\gamma(cn).$$

The equivalence class containing the growth function of a group G is independent of the generating set chosen for G, and it is called the growth of G. Exponential functions with base more than unity are equivalent, and G is said to have exponential growth if $\gamma(n) \sim 2^n$. Similarly, if $\gamma(n) \sim n^d$, G is said to have polynomial growth. In this case, d is an invariant of the equivalence class.

The growth of the fundamental group of a Riemannian manifold L determines the growth of the volume of a ball of radius n in its universal cover $\tilde{L}$; that is, it characterises the growth of $\tilde{L}$ "at infinity". More generally, the growth of a manifold Y covering a compact manifold X depends only on the fundamental groups $\pi_1(X)$, $\pi_1(Y)$ and the embedding $\pi_1(Y) \subseteq \pi_1(X)$.

Every nilpotent group (see §4 of Chapter 2) has polynomial growth, and Milnor and Wolf showed that a soluble group has polynomial growth if and only if it is nilpotent-by-finite: other soluble groups have exponential growth. The definitive result, one with no preconditions of any sort, was obtained by Gromov in 1981.

Theorem. *Every group of polynomial growth is nilpotent-by-finite.*

The idea of the proof is to introduce a "limit" of metric spaces connected with a given "abstract" group G, which then makes it possible to extend a long chain of proofs from situations involving simple restrictions on the growth of the group right up to the structural description. It goes like this.

We start with an example: $G = \mathbb{Z}^2$ is the integral lattice in the plane, $\frac{1}{2}G$ is the lattice with half-integer coordinates, *etc*. The sequence $\frac{1}{n}G$ is densely distributed in $\mathbb{R}^2$ as $n \to \infty$, and the limit of these spaces, in the precise sense to be defined below, is the whole plane $\mathbb{R}^2$.

The exact definition of the *distance* between metric spaces X an Y with distinguished points x and y is as follows. Consider all metrics δ on $X \cup Y$ whose restrictions to X and Y are their original metrics. Then the distance between (X, x) and (Y, y) is the greatest lower bound of the numbers ε over all metrics δ and all numbers $\varepsilon > 0$ with the property that $\delta(x, y) < \varepsilon$ and the ball of radius $1/\varepsilon$ centred at x is contained in the ε-neighbourhood of Y (and conversely, the $1/\varepsilon$-ball centred at y is contained in the ε-neighbourhood of X).

There is a natural left-invariant metric on G: $d(x, y)$ is the length of a shortest expression for $x^{-1}y$ in terms of the generators. The same set equipped with the metric λd given by $(\lambda d)(x, y) = \lambda d \cdot (x, y)$ is denoted by λG. For his main construction, the author first chooses a subsequence of numbers r_i such that the sequence of values $\gamma(r_i)$ of the growth function is sufficiently "regular"; using the fact that growth is polynomial he shows that the sequence (G_i, e_i), where $G_i = \frac{1}{r_i} G$, $e_i = e$, converges in the sense of the above definition of distance to a space (Y, y) with the following properties: 1) Y is homogeneous, 2) Y is connected and locally connected, 3) Y is complete, 4) Y is locally compact and finite-dimensional.

It follows from the solution of Hilbert's Fifth Problem that the group of isometries of Y is a Lie group with finitely many connected components. The rest of the proof, although admittedly subtle, is purely algebraic. It uses the theorems of Jordan and Tits on linear groups, which enable one to proceed by induction on the exponent d figuring in the polynomial growth and to bring in the above-mentioned result of Milnor and Wolf.

Despite all this, Milnor's question about the existence of a group with growth intermediate between polynomial and exponential was still unsettled. The answer was provided by Grigorchuk in 1983.

Theorem. *There exists a group G whose growth function increases faster than every polynomial but slower than every exponential*:

$$c_1 \exp n^{1/2} < \gamma(n) < c_2 \exp n^{\alpha}, \qquad \text{where} \qquad \alpha < 1.$$

An example of a group of this sort is one that we shall be looking at in more detail in section 5 of Chapter 2. The estimation of its growth is facilitated by a rewriting process (see 2.1) connected with a certain subgroup H of index 2 in G. In fact, Grigorchuk made not just one, but a wide spectrum of very diverse examples, both periodic and torsionfree, of intermediate growth. It is also shown that the set $\mathcal{R}$ of growth degrees of finitely generated groups contains a linearly ordered set with the cardinal of the continuum; nevertheless, $\mathcal{R}$ has a subset of the cardinal of the continuum consisting of pairwise incomparable growth degrees.

An important invariant of a symmetric random walk on the group $G = \langle x_1, \ldots, x_m; r_1 = 1, r_2 = 1, \ldots \rangle$ is the number $d = \overline{\lim}_{n\to\infty} \sqrt[n]{d_n}$ considered by Kesten and Grigorchuk, where d_n is the number of uncancellable words of length n contained in the kernel of the given presentation. It characterises the "growth" of this kernel. Grigorchuk showed that the group is amenable if and only if $d = 2m - 1$ (otherwise, $d < 2m - 1$). Amenable groups have applications in the theory of von Neumann algebras, dynamical systems and ergodic theory, and there are many equivalent definitions. Here is a purely combinatorial one.

Definition. A group G is said to be *amenable* if, for every finite subset A of G, every finite subset $\{a_1, \ldots, a_n\}$ of generators of G and every $\varepsilon > 0$ there is a subset B of A such that

$$|B \cap a_i B|/|B| > 1 - \varepsilon \quad \text{for} \quad i = 1, \ldots, m.$$

von Neumann proved in 1929 that the class of amenable groups is closed under the operations of taking subgroups, epimorphisms, extensions (see § 2 of Chapter 2), and unions of ascending chains. All finite and all abelian groups belong to this class. von Neumann also showed that the free group F_2 is not amenable.

The earliest examples of non-amenable groups not containing subgroups isomorphic to F_2 were discovered by Ol'shanskij as late as 1980. (The value of the parameter d for the presentations of his non-amenable groups is less than three. The construction can even be carried out for values less than any number of the form $\sqrt{3} + \varepsilon$, but further improvement is impossible because the growth of every nontrivial normal subgroup of F_2 is characterised by a number less than $\sqrt{3}$.) In 1982, Adian showed that the Burnside groups $\mathbb{B}(m, n)$ (see 5.3) are not amenable. Every non-amenable group has exponential growth, so that the Grigorchuk group of intermediate growth is amenable. It is easy to deduce from the periodicity of this group that it cannot be obtained from finite and abelian groups using the group operations mentioned in the preceding paragraph. In passing, this solves a question of Day about the structure of amenable groups.

The growth functions of various numerical characteristics of finitely generated groups and their presentations are new invariants for distinguishing between groups. Study of them and of their connections with other properties of infinite groups is currently attracting heightened attention.

Chapter 2
Structure Theory of Infinite Groups

The idea of symmetry pervades the whole of mathematics and many branches of natural science. It is formalised in the group concept. In conjunction with other concepts of a general nature, it has given birth to fundamental thrusts in the subject such as topological groups, Lie groups, linear groups and representation theory, algebraic and discrete groups, and other branches of mathematical science. It is this idea, already alluded to in the introduction, that underlies the wealth of content in group theory, with its broad concepts and diverse methods; however, it does incline one to a belief that the overall regularity in the structure of groups in the large, deep though it may be, is nevertheless unreal.

All the same, there are some unifying concepts (things like commutativity, solubility, nilpotency, simplicity, *etc*) in the language and methodology of the different branches of the theory. This unity shows up in cases where the group classes characteristic to the various branches of the theory are defined by internal, algebraic conditions. In this chapter, we select the most important abstract properties (that is, properties not involving any additional structure) of the group composition. Our task is to show that even the consideration of groups as purely algebraic objects can lead to subtle theorems about their structure.

Abelian groups are in a very special situation. Here the idea of description in terms of invariants really does find frequent incarnation. In describing other groups, the authors regard the group axioms as being sufficiently indeterminate that questions about complete classification should be illegal, even for some important group classes. Alongside theorems elucidating the internal structure of certain groups (for example, results yielding short normal series with "good" factors), we shall find room for results that reveal unforeseen connections, have clear statements or contain beautiful ideas. On the other hand, we shall eschew almost entirely results from narrow, isolated parts of the theory. (These develop when their authors have no clear goals or, disregarding the possibility of achieving what goals they have, they impose a whole set of new conditions simply in order to be able to state new mathematical propositions. This phenomenon is not confined to group theory, of course.)

§ 1. Abelian Groups

Definition. A group G is said to be *abelian* (or commutative) if $xy = yx$ for all x, y in G.

The numerous examples of abelian groups arise in the first instance from various number systems with the usual operations of addition or multiplication; however, abelian groups are also found among the principal invariants of topological spaces and complex algebraic objects. To a fair degree, their importance is reflected in surveys on problems in topology and homological algebra. For example, the homology groups $H_n(X)$ are among the most popular characteristics of a space X. They can be calculated by decomposing X into cells of different dimensions, and it is a remarkable fact that $H_n(X)$ is independent of the actual decomposition.

Knowledge of the homology groups enables one to solve such important problems as that of extending a continuous map on a subspace to the whole space, or the dual problem of lifting maps. Problems of this type are automatically insoluble if there is no solution to the parallel problems for homology groups in some dimension n. In this connection, it is appropriate to observe that the connection between algebra and topology is most often one-sided: with a topological space there is associated some algebraic object, a group for instance. This correspondence F is functorial, that is, a continuous map $X \to Y$ gives rise to a homomorphism $F(X) \to F(Y)$. (The direction of the arrows is sometimes reversed.)

For example, the related higher homotopy groups $\pi_n(X)$, $n > 1$, are abelian. ($\pi_1(X)$ is the fundamental group of X.) The fact that $G = \pi_{n-1}(B^n) = \{0\}$, while $H = \pi_{n-1}(S^{n-1}) \neq \{0\}$, where B^n is the n-dimensional ball and S^{n-1} the $(n-1)$-sphere, can be used to show that B^n cannot be mapped continuously onto its boundary S^{n-1} in such a way that each point of S^{n-1} remains fixed. (The "barrel theorem".) Otherwise, there would be a commutative diagram

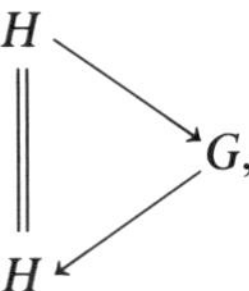

for the corresponding group homomorphisms, which is impossible since $H \neq \{0\}$, $G = \{0\}$.

Abelian group theory also has connections with the foundations of mathematics, axiomatic set theory, and ring theory; and there have been great achievements in its internal development. However, we shall restrict our attention here to the most important of the elementary questions.

It is usual to write abelian groups additively when investigating their general properties.

1.1. Finitely Generated Abelian Groups

Definition. A *free abelian group* is a direct sum $F = \sum_{i \in I}^{\oplus} A_i$ of a set of infinite cyclic groups A_i. The number $|I|$ of direct summands is called the *rank* of F.

The invariance of the rank of a free abelian group is proved in the same way as the invariance of the dimension of a vector space. If we are thinking about linear independence relative to integer coefficients, the generators of the cyclic summands are linearly independent; conversely, every group with a linearly independent generating set (a *basis*) is free abelian.

Clearly, every map of the basis $\{x_i\}_{i \in I}$ into an arbitrary abelian group has a unique extension to a homomorphism of the free abelian group defined by the formula $\phi(\sum k_i x_i) = \sum k_i \phi(x_i)$, so that every abelian group is a homomorphic image of a free abelian group: it is enough to take a free abelian group with basis big enough to map to a generating set of the given group.

Every subgroup of an n-generator abelian group is n-generator. This elementary fact is proved by induction on n, the first step being that subgroups of cyclic groups are cyclic (a fact well known from any first course on the subject).

Let G be a finitely generated abelian group. It is a factor-group F/S of a free abelian group $F = \langle x_1, \ldots, x_r \rangle$ of suitable finite rank r. We write the generators s_i, $i = 1, \ldots, r$, of S in terms of this basis: $s_i = \sum_{j=1}^{n} a_{ij} x_j$, $a_{ij} \in \mathbb{Z}$. The matrix (a_{ij}) is called the relation matrix of G. It depends on the basis $\{x_1, \ldots, x_r\}$ and the generating set $\{s_1, \ldots, s_r\}$ of S.

We shall alter the basis for F using the following elementary transformations:

1) replace x_i by x_j and x_j by x_i;
2) replace x_i by $x_i + kx_j$, $k \in \mathbb{Z}, j \neq i$;
3) replace x_i by $-x_i$.

We also perform the analogous transformations on the generating set for S. It is clear that the relation matrix also undergoes elementary transformations, on its rows and columns. Conversely, every elementary transformation on the relation matrix is caused by some transformation of the basis or the generating set of S.

Every integer matrix can be brought by elementary transformations of its rows and columns to diagonal form (this is a simple fact established by induction on the size of the matrix). Thus, a basis $\{y_1, \ldots, y_r\}$ for F and a generating set $\{t_1, \ldots, t_r\}$ for S can be chosen that are connected by relations of the form $t_i = d_i y_i$, $i = 1, \ldots, r$, $(d_i \in \mathbb{Z})$. Among other things, it follows from this that the nonzero t_i constitute a basis for S (that is, every subgroup of a free abelian group is free abelian); moreover,

$$F/S \cong \textstyle\sum^{\oplus} \langle y_i \rangle / \langle d_i y_i \rangle.$$

We have therefore proved the following fundamental theorem:

Theorem. *Every finitely generated group G can be decomposed as a direct sum of finitely many cyclic groups:* $G = \sum^{\oplus} \langle g_i \rangle$.

The number of infinite cyclic summands in this decomposition is equal to the defect of the relation matrix (a_{ij}) and is an invariant of G – the "torsionfree rank" $r_0(G)$. (In the case of the homology group $H_n(X)$ of a topological space, this number is the n-th Betti number of X.)

The fundamental theorem can be sharpened:

Theorem. *There exists a decomposition of every finitely generated group G as a direct sum of cyclic groups such that the orders $d_1, \ldots, d_k$ of the finite summands satisfy the divisibility conditions $d_i | d_{i+1}$ for $i = 1, \ldots, k - 1$. The numbers $d_1, \ldots, d_k$ are invariants of G.*

Consequently, there is a complete and explicit classification of finitely generated abelian groups. There exists a second version of the decomposition, in which all the d_i are prime-powers. In that case, it is the totality of them that is an invariant of G (the so-called elementary divisors). In conclusion, we remark that the proof given above carries over to finitely generated modules over principal ideal rings. (Abelian groups are modules over $\mathbb{Z}$.) In particular, one of the versions of the existence and uniqueness proof for Jordan forms of linear operators involves viewing a vector space as a module over the ring of polynomials in a single variable.

1.2. Divisible (Complete) Abelian Groups. The class of divisible groups has an important part to play in abelian group theory.

Definition. A *divisible* (complete) group G is one in which the equation

$$nx = g \tag{1}$$

is soluble for every integer $n \neq 0$ and every g in G.

An example is the additive group $\mathbb{Q}$ of rational numbers. Another is the multiplicative group of the p^k-th roots of unity, $k = 1, 2, \ldots$. Note that the multiplicative form of equation (1) is $x^n = g$. If $n = p^k m$ with $(m, p) = 1$, we need to observe that every element of our group has p^k-th roots, while the equation $x^m = h$ in a p-group has a unique solution whenever m is prime to p. (This last result follows from the fact that the map $y \to y^m$ is an automorphism of a cyclic p-group, which means that each of its elements can be written as y^m.) This group, which is called a group of type p^∞, or a *quasicyclic* p-group (notation: C_{p^∞}) since its lattice of subgroup is a chain (like that of a cyclic p-group), can be defined in terms of generators $a_1, a_2, \ldots$ with defining relations $pa_1 = 0$, $pa_2 = a_1, \ldots, pa_{i+1} = a_i, \ldots$.

Since $\mathbb{Z} \subseteq \mathbb{Q}$, every free abelian group can be embedded in a divisible group $D = \sum^{\oplus} Q_i$, where $Q_i \cong \mathbb{Q}$. Thus, every abelian group $A \cong F/S$ is isomorphic to a subgroup of the divisible group D/S. In fact, there is a minimal divisible group containing A as subgroup. It is unique up to isomorphism over A and is called the completion[3] of A.

The following is a more fundamental property of divisible groups.

Theorem. *Every divisible subgroup H of an abelian group G is a direct summand*: $G = H + K$.

It is easy to deduce a complete description of divisible abelian groups from this.

Definition. The *periodic part* of a group G is the set of elements of G of finite order.

The periodic part of an abelian group is always a subgroup, since $mn(a + b) = 0$ whenever $ma = 0$, $nb = 0$.

Theorem. *Every divisible abelian group is a direct sum of groups isomorphic to the additive group $\mathbb{Q}$ or to quasicyclic groups.*

For, the periodic part of a divisible group is obviously divisible; it is therefore a direct summand, so that the whole group is the direct sum of a periodic divisible group and a torsionfree divisible group. It is not difficult to see that a torsionfree divisible abelian group carries the structure of a linear space over the field of rational numbers ($\frac{m}{n}a$ is defined to be the solution of the equation $nx = ma$). It has a basis, and thus is decomposable as the direct sum of one-

[3] For the reader familiar with module terminology, we note that the completion is the injective hull of the $\mathbb{Z}$-module A, and the next theorem means that divisible groups are injective as $\mathbb{Z}$-modules.

dimensional subspaces, that is, subgroups isomorphic to $\mathbb{Q}$. We still have to consider the periodic part T. If we take any element a_1 of order p, and a_2 as any solution of the equation $px = a_1$, *etc*, we shall find a quasicyclic subgroup, so that every nontrivial periodic divisible group has quasicyclic subgroups. We consider all the subgroups that are direct sums of quasicyclic groups, and order them by stipulating that $U \preceq V$ if U is a direct summand of V. It is clear that Zorn's lemma applies to this ordered set: let T_0 be a subgroup that is maximal among those under discussion. Since it is divisible, we have $T = T_0 \oplus T_1$, where T_1 is also divisible, so that it has a quasicyclic subgroup if not trivial, and T_0 is not maximal. This means that $T = T_0$.

Every abelian group has a maximal divisible subgroup. It splits off as a direct summand, and the complementary subgroup (or factor-group by the "divisible part") is determined uniquely up to isomorphism. It is called a *reduced* group. The normal objects of study are the reduced groups, since the divisible part can be viewed as fully described.

The scheme underlying the structure theory of abelian groups can be described as follows. The factor-group of any group by its periodic part is torsionfree. Therefore we need to study periodic groups, torsionfree groups, and the rules for constructing a group when its periodic part and the factor-group by it are known. We sketch now how this scheme is realised in practice.

1.3. Periodic Abelian Groups

Definition. The p-primary component of a group for the prime p is the set of all elements of orders p^k, $k = 0, 1, \ldots$.

By the Chinese Remainder Theorem, it follows that every element of order $n = p_1^{\alpha_1} \ldots p_t^{\alpha_t}$ is a sum of parts having orders $p_1^{\alpha_1}, \ldots, p_t^{\alpha_t}$. Thus, every periodic abelian group is the direct sum of its *primary components*, and the study of periodic abelian groups is equivalent to that of p-groups. This part of the theory has been developed most fully in the case of countable groups. (Incidentally, interest in cardinality questions and axiomatic set theory is typical in abelian group theory: this or that theorem can be proved in this or that axiom system; fortunately, countable groups constitute an exception.)

Prüfer's second theorem characterizes countable p-groups expressible as direct sums of cyclic groups – this is the most widely-used and transparently-constructed class of p-groups.

Definition. An element $g \neq 0$ of a p-group G has *infinite height* if the equation $p^k x = g$ has a solution for all natural numbers k. If the equation $p^l x = g$ has a solution but $p^{l+1} x = g$ does not, then g is said to have height l.

The existence of elements of infinite height in a group does not mean that it has nonzero divisible subgroups, since the equations $px_1 = x$, $p^2 x_2 = x$ do not imply that $px_2 = x_1$, and there need be no subgroup of type p^∞ in G.

Theorem. *A countable abelian p-group can be expressed as a direct sum of cyclic subgroups if and only if it has no nontrivial elements of infinite height.*

In one direction the theorem is obvious, and is not hard to prove in the other direction. Countability is essential. A counterexample is the periodic part of the Cartesian product $\prod_{i \in \mathbb{N}} \langle a_i \rangle$ of cyclic groups $\langle a_i \rangle$ or order p^i. The proof of this fact, and of Prüfer's theorem, can be found in Kargapolov and Merzlyakov (1982), for instance. Prüfer's first theorem holds for groups of arbitrary cardinality. It states that every abelian group of finite exponent (that is, $nG = 0$ for some positive integer n) can be expressed as a direct sum of cyclic subgroups. The theorem is proved by induction on the length of the decomposition $n = p_1^{\alpha_1} \dots p_t^{\alpha_t}$. For $n = p$, it is equivalent to the fact that a linear space over $\mathbb{Z}_p$ has a basis, that is, is a direct sum of one-dimensional subspaces.

The elements of infinite height in a p-group G form a subgroup $G_1 = \bigcap_{k=1}^{\infty} p^k G$. This construction can be repeated for G_1: $G_2 = \bigcap_{k=1}^{\infty} p^k G_1$, and so on, for all natural numbers and then over the ordinals. If G is reduced, then $G_\alpha = \{0\}$ at some point. The series thus obtained is called the Ulm series, and its factors G_i/G_{i-1} the *Ulm factors*. By Prüfer's second theorem, these factors are direct sums of cyclic groups, and all except possibly the last (if there is a last one) has elements of unbounded orders. It turns out that a countable group is determined by its Ulm factors; with the exceptions of the conditions mentioned, these factors can be arbitrary. See Kurosh (1957) for more details.

Thus, we can say that countable periodic groups have a reasonably easy description. In the uncountable case, there are more negative examples than theorems. The situation here was clarified by Kulikov using the concept of basic subgroup.

Definition. A subgroup A of an abelian group G is said to be *pure* in G if, for each element a of A and each n in $\mathbb{Z}$, the equation $nx = a$ is soluble in A if it is soluble in G.

In particular, direct factors of an abelian group G are pure subgroups, as are its periodic part and indeed any subgroup A such that G/A is torsionfree. It is easy to check that a subgroup A is pure if and only if, for every prime p, the p-height of each element a of A in A is the same as its p-height in G.

Definition. A subgroup B of an abelian p-group G is said to be a *basic subgroup* if it is pure in G, is a direct sum of cyclic subgroups, and the factor-group G/B is divisible.

Theorem (Kulikov). *Every abelian p-group G has a basic subgroup B. All the basic subgroups of G are mutually isomorphic.*

It is thus the case that for general periodic groups there are important classifying subgroups (the basic subgroups), while countable p-groups are completely described. This means that the first part of our programme can be considered to be achieved. We come now to the second part.

1.4. Torsionfree Abelian Groups. Integral linear dependence provides us with the concept of the *rank* of an abelian group: the maximum number of linearly independent elements. (For finitely generated A is just the number $r_0(A)$ mentioned in 1.1). A torsionfree group G of rank r is a subgroup of the additive group of the linear space $\mathbb{Q}^r$ of dimension r over $\mathbb{Q}$. For, if $\{a_1, \dots, a_r\}$ is a maximal linearly independent set, every element g of G is linearly dependent on $a_1, \dots, a_r$, so that $kg = l_1 a_1 + \cdots + l_r a_r$, where the coefficient-row $(k, l_1, \dots, l_r)$ is unique up to proportionality. Now simply associate with g the row $\left(\frac{l_1}{k}, \dots, \frac{l_r}{k}\right)$ in the arithmetic space $\mathbb{Q}^r$ over $\mathbb{Q}$.

Let $(p_1, p_2, \dots)$ be the set of all primes. The *characteristic* of an element g of G is the sequence $(k_1, k_2, \dots)$, where k_i is a non-negative integer or ∞ according to the type of solution that exists for the equation $p_i^l x = g$: $k_i = l$ if g is divisible by p_i^l but not by p_i^{l+1}, and $k_i = \infty$ if g is divisible by every power of p_i.

Changing g to a multiple ng or to a submultiple x (so that $nx = g$ for some n) alters the characteristic of g in only finitely many places, namely those for which $k_i \neq \infty$ and p_i occurs in the canonical decomposition of n. Such characteristics are said to be equivalent; in a group of rank 1, all nonzero elements have the same characteristic. Thus, a group of rank 1 has a uniquely determined *type* – the equivalence class of the common characteristic – and the converse is also true. This description is pretty transparent; however, there are continuously may groups of rank 1 – one can construct continuously many inequivalent characteristics involving 0 and ∞ only. For example, the additive groups $H\frac{m}{2^k}$ and $H\frac{m}{3^k}$ of 2-adic and 3-adic rationals are non-isomorphic since they correspond to the types containing $(\infty, 0, 0, \dots)$ and $(0, \infty, 0, \dots)$.

For completely decomposable groups – direct sums of groups of rank 1 – the direct factors are uniquely determined up to isomorphism. In general, however, groups of rank at least 2 are very complicated. The first important example here was the group of rank 2 constructed in Pontryagin (1954) that is not decomposable as a direct sum of groups of rank 1 (Pontryagin duality theory then applies to obtain a compact connected two-dimensional abelian group not decomposable as a direct sum of one-dimensional subgroups). In fact, a group of finite rank may have two nonisomorphic decompositions as direct sums $G = G_1 \oplus G_2 = H_1 \oplus H_2$ with indecomposable summands. Finally, decompositions with different numbers of indecomposable factors are possible. One can expect arbitrarily improbable surprises with groups of infinite rank! Of the "positive" results we mention the Pontryagin freeness criterion.

Theorem. *A torsionfree group G is free if and only if all its subgroups of finite rank are free.*

Despite the unpleasantnesses enumerated above, there does exist something of a classification of torsionfree groups of finite rank. Obviously, it is going to be cruder than for groups of rank 1: in this case, the group is associated with a sequence of matrices. Nevertheless, the problem has been solved in principle in terms of divisibility (Derry, Mal'tsev), and the description obtained satisfies some very stringent conditions. All of this highlights the fact that the mere description of groups is not the sole aim of the theory, this description being not at all convenient to implement. It can be said that the problem of whether a group is directly decomposable is insoluble in terms of invariants. It is not surprising that this part of the theory continues to develop, even though new classifications of groups of finite rank have appeared.

Thus, the theory of torsionfree groups is somewhat less satisfactory than that of periodic groups. However, the third stage in our programme – extension theory – seems even darker.

1.5. Extensions in the Class of Abelian Groups. Not every abelian group is a direct sum of its periodic part and a torsionfree group. An example is the Cartesian product $G = \prod_{p \in \pi} A_p$ of reduced p-groups A_p over an infinite set π of prime numbers. Cleary, the periodic part consists of sequences for which almost all components are zero, that is, it can be identified with the direct sum $\sum_{p \in \pi}^{\oplus} A_p$. The fact that G does not split follows because the factor-group is divisible, while G has no divisible subgroups. To show that $\prod A_p/\sum^{\oplus} A_p$ is divisible, we can discard finitely many components, noting that they lie in $\sum A_p$. (More generally, $\prod A_p/\sum^{\oplus} A_p$ does not depend on any one of the groups A_p, only on the totality of them!) In solving the equation $mx = g$, we can assume that none of the primes p in π divides n. The homomorphism $x \to mx$ of each cyclic subgroup of A_p has trivial kernel because of this assumption, so that it is an automorphism. This means that every element is of the form mx, so that our equation is soluble in A_p. Thus $\prod A_p/\sum^{\oplus} A_p$ is divisible. The commutativity of the groups A_p is not essential in this argument, and we shall use this fact again later.

Nevertheless, there is an important case in which the periodic part is guaranteed to split off.

Theorem (Baer, Fomin). *Every abelian group with periodic part T splits if and only if T is a direct sum of a group of finite exponent and a divisible group.*

In the general case of non-split extensions, "factor-systems" (see 2.2) and the language of homological algebra is brought into play. However, there is a simple method, unique to abelian group theory, of introducing a group operation on the set of pairwise inequivalent extensions, using Baer sums.

Definition. An *extension E of a group A by a group B* is a short exact sequence

$$E\colon A \xrightarrow{\mu} G \xrightarrow{\varepsilon} B \tag{2}$$

in which μ is a monomorphism, ε is an epimorphism, and "exactness" means that $\mu(A) = \operatorname{Ker} \varepsilon$.

Very often, G itself is referred to as the extension, and $\mu(A)$ is identified with A. Then $G/A \cong B$.

Definition. Extension (2) is *equivalent* to the extension $E_1: A \overset{\mu_1}{\to} G_1 \overset{\varepsilon_1}{\to} B$ if there exists an isomorphism $\phi: G \to G_1$ producing the commutative diagram

$$\begin{array}{ccccc} A & \xrightarrow{\mu} & G & \xrightarrow{\varepsilon} & B \\ \| & & \downarrow{\scriptstyle\varphi} & & \| \\ A & \xrightarrow{\mu_1} & G_1 & \xrightarrow{\varepsilon_1} & B \end{array}.$$

If $\alpha: A \to A'$ is any epimorphism, there is an extension αE arising from (2) (it can be interpreted as the quotient of G by $\operatorname{Ker} \varepsilon$ – see Fuchs (1973) for more details). Dually, $E\beta$ is defined for any monomorphism $\beta: B' \to B$. If we now look at the direct sum

$$E \oplus E_1: A \oplus A \xrightarrow{\mu \oplus \mu_1} G \oplus G_1 \xrightarrow{\varepsilon \oplus \varepsilon_1} B \oplus B,$$

and take β to be the diagonal embedding of B in $B \oplus B$ ($\beta(b) = (b, b)$), and α to be the codiagonal epimorphism from $A \oplus A$ to A (that is, $\alpha(a_1, a_2) = a_1 + a_2$), then the extension $\alpha(E \oplus E_1)\beta$ of A by B is called the *sum* $E + E_1$ of the extensions E and E_1.

The operation of *addition* of extensions thus defined is consonant with the equivalence relation on extensions, and the resultant addition of equivalence classes of extensions satisfies the axioms for an abelian group. It is denoted by $\operatorname{Ext}(B, A)$; the zero is the class of the split extension, that where $\mu(A)$ is a direct summand of G.

It follows from 1.2 that $\operatorname{Ext}(B, A) = \{0\}$ if A is divisible. The same is true if B is free. Somewhat less obvious is the isomorphism $\operatorname{Ext}(\mathbb{Z}_n, A) \cong A/nA$ for arbitrary A and all integers n. In calculating extension groups one uses such general properties as

$$\operatorname{Ext}\left(\sum_i{}^{\oplus} B_i, A\right) \cong \prod_i \operatorname{Ext}(B_i, A), \qquad \operatorname{Ext}\left(B, \prod_i A_i\right) \cong \prod_i \operatorname{Ext}(B, A_i).$$

By 1.1, these properties are sufficient for a determination of $\operatorname{Ext}(B, A)$ when B is finitely generated. In more general situations there re no such prescriptions. Furthermore, in most cases interest centres on determining the extension group itself, and little is known about properties of the group elements or indeed the properties of the extensions. In fact, it would be astonishing were the situation otherwise. $\operatorname{Ext}(B, A)$ has been described in many cases by Kulikov in terms of products similar to Cartesian products.

We conclude our discussion of abelian group extensions by stating a result that is valid only in the abelian case.

Theorem. *Let A be a subgroup of an abelian group B and $\mu: B/A \to C$ a monomorphism into some abelian group C. There exists an abelian group D containing B and an extension of μ to an isomorphism $D/A \cong C$.*

1.6. Algebraically Compact Groups. Having described our investigative scheme, we single out in conclusion just one special class of abelian groups for discussion.

The fundamental results of Pontryagin (1954) in the theory of topological groups pointed up the importance of the role of compact abelian groups; these are connected with discrete groups by the duality theorems, and character theory for compact groups has been in the classical domain for some time now. The study of the abstract structure of compact abelian groups led Łos and Kaplansky to the isolation of algebraically compact groups. This class has a number of remarkable algebraic properties, and it is one of the most important objects in the transition to abelian groups of uncountable cardinality.

Definition. A group A is said to be algebraically compact if it is a direct summand of every group G containing it as a pure subgroup.

As we showed in 1.2, all divisible groups are algebraically compact. Those are the obvious examples, and an arbitrary group is algebraically compact if and only if its reduced part is. Another example is the additive group of the p-adic integers. However, the additive group of integers (and indeed every free abelian group of rank at least 1) fails to be algebraically compact. By the theorem in 1.5, groups of finite exponent are algebraically compact. Finally, injective modules over an arbitrary ring are algebraically compact *qua* abelian groups.

Homomorphisms $A \to B$ of abelian groups can be added as functions. The group $\mathrm{Hom}(A, B)$ so obtained is algebraically compact if A is periodic or B is algebraically compact (Fuchs).

The next result reflects the multi-faceted nature of this concept.

Theorem. *The following conditions on an abelian group A are equivalent*:

1) *A is algebraically compact*;

2) *A is algebraically a direct summand of a compact topological group*;

3) *if every finite subsystem of any system of equations over A has a solution in A, the system itself has a solution in A*;

4) *A is a direct summand of a Cartesian product of finite cyclic and quasicyclic groups.*

The instruments for studying algebraically compact groups are two naturally-occurring subgroup topologies on them: the $\mathbb{Z}$-adic and the p-adic. For the first of these, a base of neighbourhoods of the identity of A is the set of subgroups nA, $n \in \mathbb{Z}$; in the second case, the subgroups $p^k A$, $k = 1, 2, \ldots$.

Theorem. *A group A is a reduced algebraically compact group if and only if it is complete in the $\mathbb{Z}$-adic topology.*

The completeness condition here is expressed by the facts that $\bigcap_{n=1}^{\infty} nA = \{0\}$, and, for every sequence of elements a_i of A such that

$$(\forall n \exists N \forall i, j \geqslant N)(a_i - a_j \in nA)$$

there exists an element a of A such that $(\forall n \exists N \forall i > N)(a_i - a \in nA)$.

Ershov has re-proved this result using a lemma stating the following: if A is a pure subgroup of B, there exists an extension C of B such that A is pure in C and C/A is divisible. He also emphasised the importance of algebraically compact groups in module theory, by observing that every abelian group A is elementarily equivalent to an algebraically compact group B. (In other words, the sets of closed formulae in first order predicate calculus valid in A and in B are the same.)

§ 2. Group Extensions and Homological Questions

The problem in extension theory is the study of groups G having a given normal subgroup A with given factor-group G/A. Clearly, to describe all finite groups it is enough in principle to solve this problem and to classify all simple groups. While decisive progress been made on the latter question, the general extension problem seems inaccessible even for finite groups. As confirmation of this remark, it is enough to recall that finite p-groups seems exceedingly amorphous objects at the moment, despite the fact that every finite p-group has a normal series (see § 3) whose factors are of the very simplest sort, namely isomorphs of the residue group $\mathbb{Z}_p$.

Infinite groups usually do not have finite normal series with simple factors. However, the structure of a group is significantly simpler if there is a "short" series (say, of length 2 or 3), one of whose factors is finite cyclic, another abelian or nilpotent, and the third perhaps a concretely given group. As we shall see, such a situation is fairly typical in the structure theory of infinite groups. The extension problem is thus urgent here too. At the same time, it is hopelessly difficult in situations that are too general, and in this section we shall select a few important special cases and some in "general position", for instance in connection with cohomology.

The homology and cohomology groups of many groups arose even earlier than the groups themselves, in the guise of (co)homology groups of topological spaces. However, their applications in general group theory were very restricted until quite recently. Currently, new connections between the internal and cohomological properties of groups are being discovered. Certain invariants, such as the cohomological dimension, have turned out to be natural from other points of view. The hope has thus been conceived of extending the potential applications of homological language even to cases where the statement of a question, and its answer, do not involve homological terminology.

2.1. Semidirect Products and Wreath Products. The notation $A \lhd G$ will always mean that A is a normal subgroup of G.

Definition. A group G is said to be a semidirect product of its subgroups A and B (in symbols: $G = A \rtimes B$) if $G = AB$, $A \lhd G$ and $A \cap B = \{1\}$. (This last condition means that expressions of the form $g = ab$, $a \in A$, $b \in B$, are unique.)

This term is more characteristic of noncommutative groups, since in the abelian case we have that G is the direct product $A \times B$. In general, the semidirect product is direct if B is normal in G. (Because, for a in A and b in B, the commutator $aba^{-1}b^{-1} = a(ba^{-1}b^{-1}) = (aba^{-1})b^{-1}$ is in the intersection $A \cap B = 1$, so that A and B commute elementwise.)

In the case of a semidirect product, we also say that G is a *split extension* of A by B, since $G/A = AB/A \cong B/A \cap B \cong B$, that is, the factor-group appears in G itself and we are dealing with the simplest type of extension.

Semidirect products occur fairly frequently. As an example, the group $T_n(k)$ of all nonsingular upper-triangular matrices over a field k is the semidirect product of the normal subgroup $UT_n(k)$ consisting of the unitriangular matrices (those with the identity down the main diagonal) and the subgroup $D_n(k)$ consisting of all the diagonal matrices.

Of course, A and B do not determine G uniquely. For instance, it is easy to see from the presentation for S_3 discussed in section 2.1 of Chapter 1 that $S_3 = \langle c \rangle_3 \leftthreetimes \langle a \rangle_2$ is the semidirect product of the cyclic groups of order 3 and 2, and a and c do not commute. On the other hand, the direct product $\langle c \rangle_3 \times \langle a \rangle_2 \cong \mathbb{Z}_6$ is abelian.

To be able to determine a semidirect product uniquely, note first that on multiplying two elements $a_1 b_1$ and $a_2 b_2$ we get $(a_1 b_1 a_2 b_1^{-1})(b_1 b_2)$, where $a_1 b_1 a_2 b_1^{-1} \in A$ since A is normal, and $b_1 b_2 \in B$. Therefore, in addition to the group operations in A and B, one must how the elements of B "act" on the normal subgroup A by conjugation. The term "action" indicates that the map α_b: $a \to bab^{-1}$ is an automorphism of A; it is very easy to check that the map ϕ: $B \to \operatorname{Aut} A$ given by $\phi(b) = \alpha_b$ is a homomorphism.

In this way, a semidirect product is uniquely determined by the triple (A, B, ϕ), where ϕ is a homomorphism from B to the automorphism group of A. Conversely, if groups A and B and a homomorphism ϕ: $B \to \operatorname{Aut} A$ are given, one can always construct a semidirect product as the set of all pairs (a, b) with the multiplication rule

$$(a_1, b_1)(a_2, b_2) = (a_1 a_2^{\varphi(b_1)}, b_1 b_2). \tag{1}$$

Here A (B respectively) is identified with the set of pairs of the form $(a, 1)$ (pairs $(1, b)$ respectively). It is not difficult to see that if we construct ϕ from a given semidirect product $G = A \leftthreetimes B$, we can recover G using formula (1) (more exactly, a group isomorphic with G). The direct product is obtained when $\phi(B) = \{1\}$.

Formula (1) can be used for constructing new groups. In particular, given an automorphism α of any group A, the homomorphism from the infinite cyclic group $\langle b \rangle$ to $\operatorname{Aut} A$ defined by $b \mapsto \alpha$ leads to the extension of A by the automorphism α. Another example: taking B to be the whole of the automorphism group $\operatorname{Aut} A$ and ϕ to be the identity map, we get the *holomorph* $A \leftthreetimes \operatorname{Aut} A$ of A.

At this point we define another of the fundamental constructs in group theory, the *wreath product*. Let A^B be the set of functions on B with values in A and componentwise multiplication: $(f_1 f_2)(x) = f_1(x) f_2(x)$ (this group is a Cartesian power of A). The group B acts on A^B by left translations α_b: $f \to f'$, where $f'(x) = f(xb)$, and at the same time we have defined a homomorphism ϕ_0: $A \to \operatorname{Aut} A^B$.

Definition. The semidirect product of A^B and B given by the homomorphism ϕ is called the unrestricted wreath product $A \mathbin{\bar{\wr}} B$ of the groups A and B.

Thus, $A \mathbin{\bar{\wr}} B$ consists of all products fb ($f \in A^B$, $b \in B$) with group multiplication $(f_1 b_1)(f_2 b_2) = (f_1 f_2^{b_1})(b_1 b_2)$, where $f_2^{b_1}(x) = f_2(xb_1)$.

The wreath product construction was born as the so-called monomial linear representation. For example, suppose that the finite group B acts regularly on the summands of the direct decomposition $V = V_1 \oplus \cdots \oplus V_m$ of a linear space V, and that the group A consists of linear operators acting on V_1 and behaving like the identity on $V_2, \ldots, V_m$. Then A and B generate their wreath product in $GL(V)$. In this form, the construction was known by Frobenius. The general definition is due to Kaluzhnin and Krasner, who established the following universal property of wreath products.

Theorem. *Every extension G of a group A by a group B can be embedded isomorphically in the unrestricted wreath product $A \mathbin{\bar{\wr}} B$.*

It can be shown that $G \cap A^B = A$ and $GA^B = A \mathbin{\bar{\wr}} B$ for the Kaluzhnin-Krasner embedding. With a choice of coset representatives $\bar{g}$ of the elements of G modulo A (see 1.1.2), the Kaluzhnin-Krasner embedding κ is given by the formula $\kappa(g) = f_g \cdot \varepsilon(g)$, where $\varepsilon(g)$ is the canonical image of the element g of G in $G/A = B$, and $f_g(x) = \bar{x}^{-1} g \overline{\varepsilon(g^{-1})x}$. This embedding κ is essentially independent of the choice of coset representatives – the image $\kappa(G)$ is replaced by a conjugate in $A \mathbin{\bar{\wr}} B$. The universal property of the wreath product is especially frequently used in the study of identities in group extensions (see H. Neumann (1967)).

If the set $A^{(B)}$ of functions with finite supports (that is, the functions where almost all values are 1) is taken in the definition instead of A^B, the semidirect product of $A^{(B)}$ by B constructed in the analogous way gives a group $A \wr B$ called the *restricted wreath product* of A by B. (The subgroup $A^{(B)}$ is the restricted direct product of isomorphic copies $A(b)$ of A, $b \in B$). Of course, for finite B the restricted and unrestricted wreath products are the same.

In particular, the group of triangular matrices $\begin{pmatrix} g & u \\ 0 & 1 \end{pmatrix}$ discussed in section 4.3 of Chapter 1 is the restricted wreath product of a free abelian group of rank $|X|$ by the factor-group $G = F/N$. (In this case, the subgroup $A^{(G)}$ is simply a free $\mathbb{Z}[G]$-module of rank $|X|$.) Shmel'kin has generalised the concept of wreath product, and also the Magnus embedding (see 1.4.3) in terms of his new concept. He also introduced the idea of verbal wreath products, which are widely used in the study of free groups in product varieties (H. Neumann (1967)).

2.2. Factor-Systems and the Second Cohomology Group. We consider now an arbitrary (not necessarily split) extension: $A \lhd G$ and $G/A = B$. In this case, the elements b of B can be put in one-to-one correspondence with coset representatives g_b in G, and every element of G can be written uniquely in the form ag_b, $a \in A$. Since the representatives g_b do not constitute a subgroup of G in general, a correcting factor is needed when they are multiplied:

$$g_{b_1}g_{b_2} = g_{b_1b_2}a_{b_1,b_2}, \qquad a_{b_1,b_2} \in A. \tag{2}$$

For the same reason, firstly we have that the map ϕ: $b \to \alpha_{g_b}$ (see 2.1) depends on the choice of representatives; secondly, it is not always a homomorphism to Aut A, it simply defines a homomorphism from B to the outer automorphism group of A, that is, the factor-group of Aut A by the group of inner automorphisms (those arising from conjugation by elements of A).

For these reasons, general extension theory is extremely complicated. It is simplified somewhat in the case where A is abelian and normal (a case which is important in applications), since the action of G on A by conjugation gives rise to a homomorphism ϕ: $B \to$ Aut A, as explained in section 4.3 of Chapter 1. This means that, compared with the situations with splitting extensions (section 2.1), additional information is required to prescribe G, namely the function $f(u, v)$ of two variables from B with values in A given by $f(b_1, b_2) = a_{b_1,b_2}$, where this element is as in (2). This function (or the set of values $\{a_{b_1,b_2}\}$) is called a *factor-system* of B with values in A.

Although the function f depends on the choice of coset representatives g_b, it is nevertheless not perfectly arbitrary, since a simple calculation using the associative law in G shows that

$$u \circ f(v, w) + f(u, vw) = f(u, v) + f(uv, w), \tag{3}$$

for $u, v, w \in B$, where we have written A in additive notation, and $\circ$ is scalar multiplication in the $\mathbb{Z}[B]$-module A (see section 4.3 of Chapter 1).

All solutions of the functional equation (3) for f are called 2-*cocycles* of B with coefficients in the module A. By the linearity and homogeneity of (3), they comprise a group $Z_\phi^2(B, A)$ under addition.

Every change in representatives from g_b to g_b' defines a function h: $B \to A$, given by $g_b' = h(b)g_b$. Substituting this equation in (3), we get a new factor-system $f'(u, v)$ differing from the old one by the function δh: $f' = \delta h + f$, where

$$(\delta h)(u, v) = u \circ h(v) - h(uv) + h(v). \tag{4}$$

Functions δh with property (4) are called 2-*coboundaries* of B with coefficients in A. It is immediately verified that (4) implies (3), so that the coboundaries form a subgroup $B_\phi^2(B, A)$ of $Z_\phi^2(B, A)$. This suggests that the extension of the abelian group A, together with B and the homomorphism ϕ, is completely determined by a coset of $Z_\phi^2(B, A)$ modulo $B_\phi^2(B, A)$.

The factor-group $H_\phi^2(B, A) = Z_\phi^2(B, A)/B_\phi^2(B, A)$ is called the *second cohomology group* of B with coefficients in the $\mathbb{Z}[B]$-module A. (We stress that the module structure on A is completly determined by the homomorphism ϕ.) If we now define equivalence of extensions just as we did for abelian groups, a simple calculation shows that for given ϕ the classes of equivalent extensions of an abelian group A by an arbitrary group B are in bijective correspondence with the elements of $H_\phi^2(B, A)$. It is enough to check that, for any 2-cocycle f, the pairs of the form (a, b), $a \in A$, $b \in A$, form a group under the multiplication $(a_1, b_1)(a_2, b_2) = (a_1 a_2^{\phi(b)} f(b_1, b_2), b_1 b_2)$.

If $H^2_\phi(B, A) = \{0\}$, then every extension of A by B (with fixed module action) splits. This will be the case if, for instance, A is injective in the category of $\mathbb{Z}[B]$-modules. Now $\mathbb{Z}_p[B]$-modules and $\mathbb{Q}[B]$-modules are injective for finite B provided that p is relatively prime to $|B|$, and it is this fact that lies at the heart of the result that every extension of A by B splits if $(|A|, |B|) = 1$ (this is the Schur-Zassenhaus theorem), as well as the fact that extensions of a torsionfree divisible (nilpotent) group by a finite group split.

2.3. Definition of Homology and Cohomology Groups. We make a small formal digression at this point in order to define the co(homology) groups in arbitrary dimension. To this end, we construct a free (or projective) resolution. It starts at the trivial $\mathbb{Z}[G]$-module $\mathbb{Z}$ with the augmentation homomorphism $\varepsilon\colon \mathbb{Z}[G] \to \mathbb{Z}$ (see section 4.3 of Chapter 1). The kernel of this homomorphism is, like every module, a homomorphic image of a free module P_1: $\operatorname{Im} \partial_1 = \operatorname{Ker} \varepsilon$. Similarly, there is a homomorphism ∂_2 of some free module P_2 such that $\operatorname{Im} \partial_2 = \operatorname{Ker} \partial_1$, *etc*. The complex

$$\cdots \xrightarrow{\partial_{n+1}} P_n \xrightarrow{\partial_n} \cdots \xrightarrow{\partial_2} P_1 \xrightarrow{\partial_1} P_0 = \mathbb{Z}[G] \xrightarrow{\partial_0=\varepsilon} \mathbb{Z} \longrightarrow \{0\} \tag{5}$$

of modules and homomorphisms is called a *free resolution* of the trivial module $\mathbb{Z}$. Other constructions are brought in for these resolutions (see Gruenberg (1970), for example), and not simply considerations relating to their existence.

If all terms of sequence (5) are tensored with $\mathbb{Z}$ over $\mathbb{Z}[G]$, its exactness (embodied in the equations $\operatorname{Ker} \partial_n = \operatorname{Im} \partial_{n+1}$) is lost: we have $(\operatorname{Ker} \partial_n \otimes 1) \supseteq \operatorname{Im}(\partial_{n+1} \otimes 1)$. The factor-group $H_n(G) = \operatorname{Ker}(\partial_n \otimes 1)/\operatorname{Im}(\partial_{n+1} \otimes 1)$ measures inexactness and is called the n-th *homology group* of G. For any $\mathbb{Z}[G]$-module A, the homomorphisms $P_i \to A$ form a group $\operatorname{Hom}(P_i, A)$ under addition, and this gives rise to maps $\partial_i^*\colon \operatorname{Hom}(P_{i-1}, A) \to \operatorname{Hom}(P_i, A)$ according to the rule: $(\partial_i^* f)(x) = f(\partial_i x)$. The groups $\operatorname{Ker} \partial_{i+1}^*/\operatorname{Im} \partial_i^* = H^n(G, A)$ measuring inexactness in the resulting complex are called the *cohomology groups* of G with coefficients in A. When $A = \mathbb{Z}$, the trivial $\mathbb{Z}[G]$-module, we also write $H^n(G, \mathbb{Z}) = H^n(G)$.

It is important that homology and cohomology groups do not depend on the free resolution, so that suitable resolutions can be used for concrete groups. For example, to compute the cohomology groups $H^n(\langle a\rangle_m)$ of a cyclic group $\langle a\rangle_m$ of order m, we construct a sequence

$$\xrightarrow{\partial_3} \mathbb{Z}[\langle a\rangle_m] \xrightarrow{\partial_2} \mathbb{Z}[\langle a_m\rangle] \xrightarrow{\partial_1} \mathbb{Z}[\langle a\rangle_m] \xrightarrow{\varepsilon} \mathbb{Z} \to \{0\}, \tag{6}$$

where $\partial_i(x) = (1 - a)x$ for odd i, and $\partial_i(x) = (1 + a + \cdots + a^{m-1})x$ for even i. Then $\partial_i\partial_{i+1} = 0$, since $(1 - a)((1 + a + \cdots + a^{m-1}) = 1 - a^m = 0$, and the exactness of the complex (6) is easily verified. It is also clear that $\operatorname{Hom}(\mathbb{Z}[\langle a\rangle_m], \mathbb{Z}) \cong \mathbb{Z}$, since every homomorphism is uniquely determined by the image of the identity. This means that $\partial_{2k+1}^*(x) = 0$ and $\partial_{2k}^*(x) = mx$. Thus $H^0(\langle a\rangle_m) \cong \mathbb{Z}$, $H^{2k}(\langle a\rangle_m) \cong Z_m$ for $k > 0$, and $H^{2k+1}(\langle a\rangle_m) = \{0\}$.

The role of the second cohomology group for the classification of extensions with abelian kernels is obvious from 2.2. The following result holds for the first cohomology group $H^1(B, A)$.

Theorem. *The conjugacy classes of semidirect complements of the normal abelian subgroup A in the group* $G = A \rightthreetimes B$ *are in bijective correspondence with the elements of* $H^1(B, A)$ *(the* $\mathbb{Z}[B]$*-module structure of A is defined by the action of B on A induced by conjugation in G).*

In particular, if $H^1(B, A) = \{0\}$, all semidirect complements of A in G are conjugate. This holds, for example, if A and B are finite groups of coprime orders.

For homology groups, the isomorphisms $H_0(G) \cong \mathbb{Z}$ and $H_1(G) \cong G/[G, G]$ are always valid, where $[G, G]$ is the commutator subgroup of G. When G is the fundamental group of a path-connected space X, the group $H_2(G)$ arose in the work of Hopf as the cokernel of a homomorphism from the homotopy group $\pi_2(X)$ into the homology group $H_2 X$. The second cohomology group of an arbitrary group G is also called the Schur multiplier, and it has the following description: if G has presentation $G = F/N$, then $H_2(G) \cong (N \cap [F, F])/[N, F]$.

The second homology group $H_2(G)$ has connections with the universal central extension $F/[N, F]$ of G and the number of defining relations for G. For example, if G is a group defined by a single relation $r = 1$, then N is generated by all the conjugates uru^{-1}, $u \in F$, and so it follows that the central subgroup $N/[N, F]$ is generated by the single coset $r[N, F]$, which means that the Schur multiplier of G is cyclic. Similarly, if the minimum number of generators of $H_2(G)$ is k, then G cannot be defined by fewer than k relations.

If N is factored by the smaller subgroup $[N, N]$, one obtains the (usually non-trivial) relation G-module $\overline{N} = N/[N, N]$ (see 4.3 in Chapter 1). The number of generators of this module also does not exceed the number of defining relations $r_1 = 1$, $r_2 = 1, \ldots$ of G: as G-module, $\overline{N}$ is generated by the cosets $r_1[N, N], r_2[N, N], \ldots$. It is not known whether, conversely, finite generation of $\overline{N}$ implies that G is finitely presented. Bieri and Strebel have proved the following theorem for metabelian groups (groups with abelian commutator subgroups – see 3.4).

Theorem. *If the relation module of a metabelian group G is finitely generated, then G is finitely presented.*

The set of groups $H^n(G)$ and $H_n(G)$ store a large amount of information about G. Culler has proved the following:

Theorem. *If* $f\colon G_1 \to G_2$ *is a homomorphism of finite groups inducing isomorphisms* $f_n\colon H_n(G_1) \to H_n(G_2)$ *for each n, then f is an isomorphism.*

For nilpotent groups, it is enough that f_1 and f_2 be isomorphisms.

2.4. Cohomological Dimension and Other Invariants

Definition. If there is a resolution (5) and an integer n such that $P_i = \{0\}$ for $i > n$, the smallest number n with this property is called the cohomological dimension of G and is denoted by $cd(G)$.

Equivalently, we require that $H^i(G, A) = \{0\}$ for all $i > n$ and every $\mathbb{Z}[G]$-module A. If $H \subseteq G$, then $cd(H) \leqslant cd(G)$ so that a group of finite cohomological dimension cannot contain a nontrivial finite cyclic subgroup (*cf* 2.3), that is, it is torsionfree.

Clearly, $cd(G) = 0$ if and only if $G = \{1\}$. It is also easy to check that $cd(G) = 1$ if G is free. There is another version of this: $cd(G) = 1$ if and only if every exact sequence of the form $1 \to A \to X \to G \to 1$ splits, where A is any abelian group. Free groups have this property even without the assumption that A be abelian. (In this stronger version, it is clear from the Nielsen-Schreier theorem (see 1.1.2) that the group G in a sequence $1 \to N \to F \to G \to 1$ must be free, since the sequence splits.) The very natural problem of determining the groups of cohomological dimension 1 was solved by Stallings in the finitely generated case, and Swan extended his result to arbitrary groups.

Theorem. $cd(G) = 1$ *if and only if* G *is free.*

Stallings also solved a problem of Serre by showing that every finitely generated torsionfree free-by-finite group is free. Bipolar structures (see 1.3.3) had an important part to play in the Stallings proof. His argument is based on the idea of the number of ends of a group, borrowed from topology. It can be defined in terms of the graph Γ of a finitely generated group G. The number k of infinite connected components of the graph obtained from Γ by removing finitely many edges is calculated, and the number of ends of G is the least upper bound of the numbers k arising in this way. It turns out that the number of ends of a finitely generated infinite group G is independent of how G is defined, and that it can take only the three values 1, 2 or ∞ (Oxley and Houghton). Using the first cohomology group with coefficients in $\mathbb{Z}_2[G]$, these cases can be distinguished as follows: G has 1, 2 or infinitely many ends according as to whether $H^1(G, \mathbb{Z}_2[G]) = 0$, $H^1(G, \mathbb{Z}_2[G]) \cong \mathbb{Z}_2$ or $H^1(G, \mathbb{Z}_2[G])$ has $\mathbb{Z}_2$-rank greater than one. It is obvious that the infinite cyclic group has two ends. In fact, this group and its finite extensions exhaust the class of groups with 2 ends.

Among the groups of cohomological dimension 2 there occur, in particular, all fundamental groups of nontrivial knots and the fundamental groups of two-dimensional closed surfaces other than $\mathbb{S}^2$ and $\mathbb{P}^2$. The most general result here is due to Lyndon.

Theorem. *If* G *is a torsionfree one-relator group* (*see* 1.4.2), *then* $cd(G) \leqslant 2$.

The class of groups of cohomological dimension 2 is not easy to visualise. For example, it contains continuously many 2-generator torsionfree groups in which all proper subgroups are cyclic.

The cohomological dimensions of groups acting on trees – see 1.3.4 – are amenable to calculation (Guildenhuys). In particular, for a free product with amalgamation we have

$$cd(G_1 *_A G_2) \leqslant \max(cd(G_1), cd(G_2), cd(A) + 1).$$

For every exact sequence $1 \to A \to G \to B \to 1$, $cd(G) \leqslant cd(A) + cd(B)$. For soluble groups, the inequality $cd(G) < \infty$ turns out to be connected with other finiteness conditions. Using the derived series (see 3), we can calculate the Hirsch number $\sum_i r_0(G^{(i-1)}/G^{(i)})$, with the rank r_0 as defined in 1.4. A soluble group G has finite cohomological dimension if and only if G is torsionfree and has finite Hirsch number. In this situaton, $cd(G)$ is the Hirsch number or 1 more than it (Fel'dman).

The absence of torsion is sometimes an inconvenient restriction, and the following theorem of Serre enables us to define the "*virtual cohomological dimension*" $vcd(G)$.

Theorem. *If H is a subgroup of finite index in a torsionfree group G, then $cdH = cdG$.*

Because of this result, if G is a group with torsion (so that $cdG = \infty$) having a torsionfree subgroup H of finite index, we can set $vcdG = cdH$; this definition is independent of the choice of H since $cdH = cd(H \cap H') = cdH'$ for any other torsionfree subgroup H' of finite index. In particular, $vcdG$ is defined for all polycyclic groups (see 3.2) and for many linear groups.

Using homology groups, Brown (1982) defines the Euler characteristic of a group G by analogy with that for topological spaces: $\chi(G) = \sum_i (-1)^i r_0(H_i G)$. Of course, this expression makes sense only when all summands are finite and almost all are zero (that is, G is of finite homology type). For finitely generated free abelian (or free nilpotent) groups G we have $\chi(G) = 0$, while for the (absolutely) free group F_n of rank n, $\chi(F_n) = 1 - n$. There is a simple connection with subgroups of finite index here also: if $|G:H| < \infty$, one has $\chi(H) = |G:H|\chi(G)$. This fact has enabled Wall to extend the definition of χ to certain types of group G with torsion having subroups H of finite index and for which $\chi(H)$ is defined by the above formula. In this case, $\chi(G) = \chi(H)/|G:H|$. The Euler characteristic is multiplicative on exact sequences $1 \to A \to G \to B \to 1$, that is, $\chi(G) = \chi(A)\chi(B)$ in such cases. Clearly, $\chi(G) = |G|^{-1}$ for finite G. The property $\chi(G) = \chi(G_1) + \chi(G_2) - \chi(H)$ for a generalised free product $G = G_1 *_H G_2$ gives that $\chi(SL_2(\mathbb{Z})) = \frac{1}{4} + \frac{1}{6} - \frac{1}{2} = -\frac{1}{12}$, since

$$SL_2(\mathbb{Z}) = \langle a, b; a^4 = 1, b^6 = 1, a^2 = b^3 \rangle$$

is a free product with amalgamation, as is clear from the description of $\mathrm{PSL}_2(\mathbb{Z})$ in § 3.2 of Chapter 1. The Euler characteristic also has close connections with properties formulated in standard group-theoretical language. The prime divisors p of the denominator of $\chi(G)$ are exactly those for which G has p-torsion. The following result of Brown answers a question of Serre.

Theorem. *Let G be a group of finite homology type, and m the least common multiple of the orders of finite subgroups of G. Then $m\chi(G)$ is an integer. In particular, if p^k divides the denominator of $\chi(G)$, G has a subgroup of order p^k.*

If $|G| < \infty$, this result is nothing but Sylow's theorem. Another consequence is that a sequence $1 \to A \to G \to B \to 1$ splits if A is torsionfree, B is a p-group and $p \nmid \chi(A)$.

§ 3. Soluble Groups

3.1. General Remarks. We shall first give the necessary definitions.

Definitions.

1) A series

$$G = A_0 \supseteq A_1 \supseteq \cdots \supseteq A_{k-1} \supseteq A_k = \{1\} \tag{1}$$

of subgroups of a group is said to be a normal series if each term is normal in G; a subnormal series if every term is normal in its predecessor.

2) A group G is said to be soluble if it has a subnormal series (1) with abelian factors (the factors are the factor-groups A_{i-1}/A_i, and the number k of factors is called the length of the series).

Clearly, $[A_{i-1}, A_{i-1}] \subseteq A_i$ for such a series; thus, in a soluble group the derived series (with terms defined inductively: $G' = [G, G]$, ..., $G^{(i)} = [G^{(i-1)}, G^{(i-1)}]$) reaches down to the identity, and is the shortest normal series with abelian factors. Its length is known as the derived length of G.

In Galois theory, the concept of finite soluble groups is associated with the solubility of algebraic equations in radicals. Although (infinite) soluble groups are linked in Picard-Vissiot theory in the same sort of way with solubility of differential equations in quadratures, this motivation cannot be regarded as a reason for any extensive investigations on soluble groups. Other than the impact of the theory of linear groups and Lie groups, it is internal reasons that have had the major part to play here. On the one hand, the specific nature of soluble groups and the possibilities for successful investigation are explained by the existence of "large" abelian subgroups. On the other hand, the scope of applications has to do with the fact that many naturally-occurring groups can be approximated (in the sense of 1.1.4) by soluble groups or have large soluble quotients, so that important invariants (for instance, the Alexander polynomial of a knot) can be computed.

The class of soluble groups is too wide for there to be significant deep results of a general nature. As an exception, one can cite a theorem of Yakovlev solving an old problem of Birkhoff that relates to the subgroup structure of a group. The background is that if $\phi: A \to B$ is an isomorphism, subgroups of the one group go to subgroups of the other in such a way that intersections and unions are preserved (here by $X \cup Y$ we understand $\langle X, Y \rangle$, the subgroup generated by

X and Y); that is, ϕ induces an isomorphism of the subgroup lattices of A and B. It is easy to see that the converse is false—groups with identical subgroup lattices are not necessarily isomorphic. For example, the symmetric group S_3 and the direct product of two groups of order 3 have isomorphic subgroup lattices. (In both groups, all proper subgroups are maximal, and there are four of them.) The question is this: is a group B soluble if there is a soluble group A with the same subgroup lattice? (In the example just mentioned, both groups are soluble, though of different derived lengths.)

Theorem. *If A and B are groups with isomorphic subgroup lattices and A is soluble, so is B.*

This means that solubility of groups is a property that can be recognized in the subgroup lattices.

The fact that a group has derived length at most n can be expressed by an identity. Let $x_1, x_2, \ldots$ be independent group variables. We set $\delta_1(x_1, x_2) = [x_1, x_2]$, $\ldots$, $\delta_i(x_1, \ldots, x_{2^i}) = [\delta_{i-1}(x_1, \ldots, x_{2^{i-1}}), \delta_{i-1}(x_{2^{i-1}+1}, \ldots, x_{2^i})]$. Then G satisfies the identical relation $\delta_i(x_1, \ldots, x_{2^i}) = 1$ if and only if G is soluble of length at most i.

3.2. Polycyclic Groups. *Polycyclic* groups are fundamental in the theory of infinite soluble groups. They are the group obtained from multiple extensions of cyclic groups, that is, groups having a subnormal series like (1) with cyclic factors A_{i-1}/A_i. In the finite case, solubility is equivalent to polycyclicity, while in general the polycyclic groups can be characterised as the soluble groups satisfying the maximum condition.

Definition. A group G satisfies the maximum condition (max) on subgroups if every ascending chain $H_1 \subset H_2 \subset \cdots$ of subgroups breaks off after finitely many steps.

Groups with max are also known as Noetherian groups. The condition is equivalent to the requirement that every subgroup be finitely generated.

Here is the definition of a concept dual to this one:

Definition. A group satisfies the minimum condition (min) on subgroups if every descending chain $H_1 \supset H_2 \supset \cdots$ of subgroups breaks off after finitely many steps.

Cyclic groups satisfy max, and it is easily checked that the class max is extension-closed. Polycyclic groups are Noetherian, therefore. On the other hand, finite generation of the factors in a soluble series (1) enables the series to be refined to a subnormal series with cyclic factors.

The number of infinite cyclic gactors in a subnormal series (1) does not depend on the choice of polycyclic series, since the Jordan-Hölder theorem means that any two subnormal series have isomorphic refinements, and refinements do not alter the number of infinite cyclic factors. This introduces another invariant, the *Hirsch number*, which is a convenient tool in inductive proofs.

The following theorems comprise an important linear characterisation of polycyclic groups.

Theorem (Mal'tsev). *Every soluble subgroup of* $\mathrm{GL}_n(\mathbb{Z})$ *is polycyclic.*

The proof rests on a remarkable fact concerning soluble linear groups.

Theorem (Kolchin, Mal'tsev). *Every soluble subgroup G of the group* $\mathrm{GL}_n(k)$ *of nonsingular matrices over an algebraically closed field k contains a triangulable subgroup H of finite index.*

Here triangulability means that all the matrices in H can be brought by a single conjugation to (upper) triangular form.

Our next theorem solves the converse problem, as posed by P. Hall.

Theorem (Auslander, Swan). *Every polycyclic group can be embedded in* $\mathrm{GL}_n(\mathbb{Z})$ *for suitable n.*

Thus, the polycyclic groups are precisely the soluble subgroups of the general linear groups $\mathrm{GL}_n(\mathbb{Z})$. The last of these theorems has been strengthened by Merzlyakov, who proved that the holomorph of every polycyclic group can be embedded in some $\mathrm{GL}_n(\mathbb{Z})$; this is a result that is of importance in the study of automorphism of polycyclic groups.

3.3. Soluble Groups of Finite Rank. Mal'tsev made the observation that many properties of soluble groups are reflected by those of the abelian subgroups. For example, he showed that a soluble group is polycyclic if all its abelian subgroups are finitely generated. He generalized the class of polycyclic groups very considerably when he introduced the concept of special rank, which in some problems is the analogue for groups of dimension in linear algebra.

Definition. A group G has special rank r if all its finitely generated subgroups are r-generator, and r is the smallest number with this property.

For example, the additive group of rational numbers is of rank 1, and more generally the special rank of a torsionfree abelian group G is the same as the rank $r_0(G)$ introduced in 1.4. For every exact sequence $1 \to A \to G \to B \to 1$ we have $r(G) \leqslant r(A) + r(B)$, so that the special rank of a polycyclic group is no more than the length of a polycyclic series (1) for it. The group $UT_n(\mathbb{Q})$ of unitriangular matrices over $\mathbb{Q}$ has finite special rank. On the whole, the class of soluble groups of finite special rank is significantly more complicated than the class of polycyclic groups. However, even here we have:

Theorem (Kargapolov). *If all abelian subgroups of a soluble group G have finite special rank, so does G.*

An important element in the proof of this deep theorem (see Kargapolov and Merzlyakov (1982)) is a lemma stating that the generating numbers of the p-subgroups of $\mathrm{GL}_n(\mathbb{Z}_{p^k})$ are bounded in terms of n only.

Since the additive group $\mathbb{Q}$ of rational numbers has special rank 1, it would seem natural that the special rank of every group with a rational series (that is, a subnormal series (1) with factors isomorphic to nonzero subgroups of $\mathbb{Q}$) of length r should be r. However, the proof of this conjecture is far from trivial. In his proof, Zaitsev first establishes that every finitely generated torsionfree soluble group has a subnormal series in which all factors are Artinian (satisfy min) or Noetherian. Using this, he strengthens an earlier theorem of Glushkov proved under a nilpotency assumption.

Theorem. *The special rank of an arbitrary group with a finite rational series is equal to the length of the series.*

The – at first sight eclectic – connection between the maximum and minimum conditions can be used to isolate a very important class of soluble groups of finite rank.

Definition. A group G is said to be a minimax group if it has a subnormal series (1) with each factor satisfying the minimum or maximum condition.

The soluble minimax groups are simply a subclass of the class of groups of finite special rank: the crucial nature of their role in the theory is demonstrated by the following result. As stated above, it was proved by Zaitsev for torsionfree groups; Robinson did it in the general case.

Theorem. *Every finitely generated soluble group of finite special rank is a minimax group.*

Kropholler has discovered a completely different characterisation of soluble minimax groups. The starting-point here is the following remark: a non-free cyclic module over the group algebra $\mathbb{Z}[A]$ of an infinite cyclic group A either has finite special rank as abelian group or contains a $\mathbb{Z}_p[A]$-free cyclic submodule for some prime p. It follows easily from this that a finitely generated metabelian group G (see 3.4) has finite special rank or else has the wreath product $\mathbf{Z}_p \wr \mathbb{Z}$ as a section (factor-group of subgroup). Kirkinskij has established a similar alternative in the more complicated case of extensions of abelian groups by polycyclic groups. However, in fullest form we have only the following result.

Theorem (Kropholler). *Every finitely generated soluble group not containing a section of the form $\mathbb{Z}_p \wr \mathbb{Z}$ for any prime p is minimax.*

In the proof, one can of course argue by induction and suppose that a group G not having a section of the form $\mathbb{Z}_p \wr \mathbb{Z}$ has an abelian normal subgroup A such that G/A is minimax. No real difficulty arises when the extension splits, since finite generation of G yields the existence of a finite generating set for A as $\mathbb{Z}[G/A]$-module. The nonsplit case leads to homological problems, and the cohomological considerations are entwined with Lyndon-Hochschild-Serre spectral sequences.

Kropholler has made a further interesting comment: every finitely generated minimax group is a product $A_1 \dots A_k$ of finitely many cyclic subgroups. As an example, the one-relator groups $G_n = \langle a, b; aba^{-1} = b^n \rangle$ are minimax (though not Artinian, and not Noetherian when $|n| > 1$), and we have $G_n = \langle b \rangle \cdot \langle a \rangle \cdot \langle b \rangle$.

3.4. Metabelian Groups and Related Groups

Definition. A soluble group of derived length 2 is said to be metabelian.

The commutator subgroup $G' = [G, G]$ of such a group is abelian by definition, and the action of G on it by conjugation (see 1.4.3) turns G' into a module over the commutative group algebra $Z[G/G']$. It is the potential for using the machinery of commutative algebra that explains the interest in metabelian groups; they are distinguished in the class of soluble groups by their "positive" qualities. Properties inherent in metabelian groups can occasionally be carried over to nearby group classes.

On the other hand, some problems in combinatorial group theory can be solved by passing to the factor-group by the second derived group. Such groups are metabelian, and in particular groups given by presentations can very often be distinguished by these factor-groups.

For instance, if π is the fundamental group of a knot K (see 1.4.3), its commutator factor-group π/π' is always infinite cyclic, and so it certainly cannot be used to distinguish inequivalent knots. However, the metabelian group $\pi/\pi'' = G$ is often studied in knot theory. Its commutator subgroup $A = \pi'/\pi''$ is a module over the group algebra $\mathbb{Z}[t]$ of the infinite cyclic group $\langle t \rangle = \pi/\pi'$, that is, a module over the ring $L = \mathbb{Z}[t, t]^{-1}$ of finite Laurent series with integer coefficients – "almost" the ring of polynomials in a single variable. Since G is finitely generated, it follows that A is a finitely generated module. (This is because "abelianising" a finitely generated group can be achieved by imposing finitely many relations of the form $[g_i, g_j] = 1$.) The annihilator of A in L is some ideal, I say. We have to note that not all the ideals of L are principal. However, because of the uniqueness of the decomposition into irreducible factors in L (Gauss's theorem), I is contained in a unique minimal ideal of L generated by a polynomial f having zero free term since it can be multiplied by the inverses in L of elements of the form t^n, $n \in \mathbb{Z}$. This polynomial f is an important invariant of π, and thus of the knot K, and is called the *Alexander polynomial.* Fox calculus (see 1.4.3 and Crowell and Fox (1963)) is used to compute it. Other polynomial invariants for knots arise as invariants in the classification of finitely generated modules over principal ideal rings; we mentioned this possibility in 1.1.

A brilliant page in the theory of soluble groups was that written by Phillip Hall. He distinguished a new class of groups – the finitely generated groups that are extensions of abelian normal subgroups by polycyclic groups (in particular by nilpotent groups). It turned out that these groups satisfy the maximal condition on normal subgroups (max-n), that is, ascending chains of normal subgroups break off. (Equivalently: every normal subgroup is the normal closure of finitely many elements).

We shall prove this theorem for the special case of finitely generated metabelian groups, using Hilbert's Basis Theorem. This states that every finitely generated commutative ring is Noetherian (that is, it satisfies the maximum condition on ideals); and we also need the trivial observation that finitely generated modules over Noetherian rings are Noetherian (with respect to submodules). Obviously, max-n holds for subgroups containing G', since G/G' is finitely generated and abelian. Normal subgroups of G contained in G' are submodules of the $\mathbb{Z}[G/G']$-module G, so that it is enough to show that this module is finitely generated. But this is so since it is generated by the commutators $[g_i, g_j]$ of the generators of G.

For the general proof, one establishes a special version of Hilbert's Basis Theorem. It should be observed too that Hall carried out numerous investigations in this class using group algebras of polycyclic groups.

The next result expresses an important property of this class, discovered by Hall. He proved it in a special case, and the general case is due to Roseblade.

Theorem. *Every finitely generated group G with an abelian normal subgroup A having polycyclic factor-group G/A is residually finite.*

The proof of this theorem, in which a version of Hilbert's Nullstellensatz plats a large part, is exceedingly complicated. The fact that polycyclic groups are residually finite can be deduced from the fact that they are embeddable in $GL_n(\mathbb{Z})$ (see 3.2).

Hall observed that these theorems are false in general if the commutator subgroup is nilpotent, even though the connections with linear groups makes it natural to suppose that groups in this class have certain advantages. (By the Kolchin-Mal'tsev theorem in 3.2, every soluble linear group has a subgroup of finite index with nilpotent commutator subgroup.) Furthermore, they are false for centre-by-metabelian groups (that is, extensions of central subgroups by metabelian groups).

Nevertheless, some clear spaces have appeared recently. Bieri and Strebel have proved that finitely presented groups with commutator subgroups nilpotent of class 2 (see § 4) satisfy max-n and are residually finite.

Wreath products and the Magnus embedding – see 1.4.3 and 2.2.1 – are very helpful in the study of metabelian groups. Baumslag and Remeslennikov have obtained the following analogue of a theorem of Higman (see 1.4.1).

Theorem. *Every finitely generated metabelian group can be embedded in a finitely presented metabelian group.*

This same result remains true under certain additional finiteness assumptions (Boler). Strebel has shown that a finitely generated centre-by-metabelian group is embeddable in a finitely presented centre-by-metabelian group if and only if it is abelian-by-polycyclic.

3.5. Generalised Solubility. Various generalisations of the concept of solubility are connected with the ideas of normal and subnormal system. A system $\{A_i\}_{i \in I}$ of subgroups containing G and the identity subgroup, closed under (infinite) unions and intersections, and such that $A_i \subseteq A_j$ or $A_j \subseteq A_i$ for all i, $j \in I$, is said to be a *normal system* if $A_i \lhd G$ for all i, and a *subnormal system* if normality holds at jumps, that is, where $A_i \subseteq A_j$ and there is no subgroup of the system contained between them, we have $A_i \lhd A_j$.

A good example of a subnormal system with abelian factors (that is, A_j/A_i is abelian if $A_i \subseteq A_j$ is a jump) is the set of all convex subgroups in a fully ordered group. A group G is said to be *fully ordered* if there is a relation $g \leqslant h$ defined between pairs of elements that is consonant with the group multiplication: $g \leqslant h \Rightarrow xg \leqslant xh$ and $gx \leqslant hx$ for all x in G. The convex subgroups are of basic importance: a subgroup is convex if it contains the segment $[g, h]$ whenever it contains the elements g, h. It is easy to see that any two convex subgroups are comparable under inclusion, and that the system of all convex subgroups is a subnormal system: if $A_i \subseteq A_j$, where A_j and A_i are adjacent, then gA_ig^{-1} is convex for all $g \in A_j$, which means that it is contained in A_i. The factors A_j/A_i do not contain proper convex subgroups, so that they are abelian, by the following result of Hölder.

Theorem. *Every fully ordered group without proper convex subgroups is abelian and isomorphic with a subgroup of the additive group of real numbers in the standard ordering.*

A group with a subnormal (normal) system is called an *RN-group* (an *RI-group*). These, and many other classes obtained by imposing additional conditions on normal and subnormal systems, were introduced by Kurosh and Chernikov. The Kurosh-Chernikov classes serve as a basis for the theory of generalised soluble groups; the answers to the general questions in this theory are largely known.

In fact, the properties *RN* and *RI* are too remote as generalisations of solubility. For example, free groups have *RI*; there is even a simple group G with a full order, so that it has *RN*. We give the example due to Chehata. For G we take the set of continuous piecewise-linear monotone increasing functions f of a real variable whose graphs have finitely many linear pieces, and for each f there is a segment $[a, b]$ such that $f(x) = x$ outside $[a, b]$. Multiplication in G is the usual composition of functions, and the order $f < g$ is determined by the existence of an x_0 such that $f(x_0) < g(x_0)$ and $f(x) \leqslant g(x)$ for $x \leqslant x_0$. (It is easy to check that this really is a full order and is consonant with composition of functions.) Diverging from our theme somewhat at this point, we note that G is an important example of a group well suited for explicit calculations. It has many interesting subgroups. One of these has a presentation with 2 relations:

$$G(2) = \langle a, b; [a, b^2a^{-1}b^{-1}] = 1, [a, b^4a^{-1}b^2] = 1 \rangle$$

(Brin, Squier). It contains no noncyclic free subgroups, but satisfies no nontrivial identities; $G(2)$ contains free abelian subgroups of infinite rank. This isolated example shows how very much simpler the structure of one-relator groups (see 1.4.2) is than that of finitely presented groups in general.

The properties RN, RI and many of their descendants are local properties in the following precise sense.

Definition. 1) A system of subgroups of a group is said to be a *local system* if every pair of subgroups in the system is contained in a third, and every element of the group is contained in one of the subgroups in the system.

2) A property is said to be local in a class of groups if every group having a local system of subgroups with the property has the property itself.

An example of a local system is that consisting of all finitely generated subgroups. Examples of local properties: local solubility, local finiteness (that is, solubility or finiteness of all finitely generated subgroups), and simplicity: freeness is not a local property.

The local theorems discovered by Mal'tsev (strictly speaking, they were stated and proved in the language of mathematical logic) Kargapolov and Merzlyakov (1982), Mal'tsev (1970)) have led to a uniform proof that many of the classes generalising solubility are local. The fact is that a wide circle of properties can be written as formulae in the language of predicate calculus, and Mal'tsev gave general conditions on these formulae ensuring that the corresponding group properties are local (indeed, this is so for other algebraic systems too).

As we have observed, many generalisations of solubility lead away from soluble groups. Thus, generalised soluble groups are studied with the addition of various supplementary restrictions of "finiteness" type (things like min or max on subgroups, finiteness of the rank, finiteness of abelian subgroups *etc*).

3.6. Identities in Soluble Groups and Algorithmic Questions. Many popular identities have been met already in this survey – the Burnside identity $x^n = 1$, commutativity, solubility of finite length, metabelianness. (See section 4 for nilpotent identities.) In addition to these basic types,there are very many identical relations that can be used to distinguish subtle group properties.

Definition. A variety of groups is the class of all groups satisfying some (finite or infinite) system of identical relations.

Every variety $\mathfrak{V}$ is characterized by its *free groups* – the largest factor-groups of (absolutely) free groups F that satisfy the system V of identities giving the variety $\mathfrak{V}$. Their role for $\mathfrak{V}$ is analogous to that of free groups in the general theory of groups.

Among the most important examples are: free abelian groups $F/[F, F]$, free Burnside groups (see 1.5.3), and the free soluble groups $F/F^{(l)}$ of length l, where $F' = [F, F]$, $F^{(i+1)} = [F^{(i)}, F^{(i)}]$ for $i \geqslant 1$.

The Magnus embedding (see 1.4.3 and 2.2.1) facilitates calculation in free soluble groups, which have many agreeable properties (residual finiteness, solubility of the algorithmic word and conjugacy problems *etc*).

As we saw in section 2.1 of this chapter, the group of matrices of the form $\begin{pmatrix} g & u \\ 0 & 1 \end{pmatrix}$, where $g \in F/N$ and u runs over a free $\mathbb{Z}[F/N]$-module, is isomorphic to the restricted wreath product of a free abelian group by F/N. For this reason, corresponding properties of wreath products are used in establishing various properties of the group $F/[N, N]$ embedded in the matrix group *via* the Magnus embedding.

When is a restricted wreath product $A \wr B$ residually finite? If and only if both groups are residually finite and B is finite or A is abelian. In fact, if $|B| < \infty$, the wreath product has a residually finite subgroup of finite index, namely that consisting of the functions $B \to A$. If A is abelian, the wreath product is residually in the set of wreath products $A \wr (B/N)$, where N is of finite index in B and $A \wr (B/N)$ is residually finite because of the previous remark. Commutativity of A arises because: if the element b of B goes to 1 under a homomorphism, then the permuting subgroups A and bAb^{-1} of $A \wr B$ go to the same subgroup, so that $[A, A]$ goes to 1.

As a corollary, we get that $F/[N, N]$ is residually finite if F/N is residually finite.

A wreath product $A \wr B$ is conjugacy separable (see 4.5) if A and B have the property, $A \wr B$ is residually finite and B has separable cyclic subgroups (Remeslennikov). In their consideration of conjugacy separability of free soluble groups, Remeslennikov and Sokolov showed that in the Magnus embedding, elements of $F/[N, N]$ are conjugate if and only if their images in the wreath product – the group of matrices of the form $\begin{pmatrix} g & u \\ 0 & 1 \end{pmatrix}$ – are conjugate, and they made the corresponding deduction about $F/[N, N]$.

These examples make direct use of properties of wreath products. There are more complicated examples. However, the advantages of the Magnus embedding for studying free groups in product varieties are quite clear.

Lichtman used the same approach in finding criteria for the groups $F/[N, N]$ to be residually nilpotent. We say that a group G is *discriminated* by the groups in a class $\mathscr{K}$ if, for every finite set $g_1, \ldots, g_n$ of elements of G there is a homomorphism $G \to C \in \mathscr{K}$ such that the images of $g_1, \ldots, g_n$ are different (G is residually in $\mathscr{K}$ if the images of two different elements are different).

A necessary and sufficient condition for $F/[N, N]$ to be residually nilpotent is that F/N be residually torsionfree nilpotent (so that, by results of Hartley, $F/[N, N]$ is residually torsionfree nilpotent) or F/N is discriminated by nilpotent p-groups of finite exponent. This is exactly the condition that the powers of the augmentation ideal (1.4.3) of the group ring $\mathbb{Z}[F/N]$ intersect in zero.

We mention next a theorem of Romanovskij, which is remarkable in that free soluble groups are applied, rather unexpectedly, to a general question in combinatorial group theory (and for which there is still no "purely" combinatorial solution).

Theorem. *Let G be the group obtained by imposing m additional relations on the free soluble group $F_n/F_n^{(l)}$ of length l and rank n, with $m < n$. Then $n - m$ generators $a_{i_1}, \ldots, a_{i_{n-m}}$ can be chosen from a free generating set $a_1, \ldots, a_n$ such that the image of $\langle a_{i_1}, \ldots, a_{i_{n-m}} \rangle$ in G is free soluble of length l and rank $n - m$.*

The following corollary solves a problem of Lyndon.

Corollary. *Suppose that $G = \langle a_1, \ldots, a_n; r_1 = 1, \ldots, r_m = 1 \rangle$, where $m < n$. Then, for some system of $n - m$ generators, the subgroup $\langle a_{i_1}, \ldots, a_{i_{n-m}} \rangle$ of G is (absolutely) free.*

The case $m = 1$ follows from the Magnus Freiheitssatz (see 1.4.2). However, as we have seen already, one-relator groups are sharply distinguished by their properties and the ease with which they can be investigated.

There is an important finiteness condition couched in terms of identities, namely the finite basis condition for the identities of a group G. It means that all the identities that hold in the given group are consequences of finitely many such identities. (The condition can be restated in the language of subgroups of the absolutely free group F of countably infinite rank invariant under all homomorphisms $F \to F$.)

Not all soluble groups have finite bases for their identities (Ol'shanskij); there is even a group with no finite basis for its identities that is an extension of a nilpotent group of class 2 by a nilpotent group of class 2 (Vaughan-Lee).

We claim that the variety $\mathfrak{B}_4\mathfrak{B}_2$ consisting of extensions of groups of exponent 4 by groups of exponent 2 does not have a finite basis for its identities. A basis for this variety is the set of words $u_n = (x_1^2 \ldots x_n^2)^4$, $n = 1, 2, \ldots$. For, if $u_n \equiv 1$ for all n, the subgroup generated by the squares has exponent 4. The words u_n are not independent; if we set $x_n = 1$, we get u_{n-1}, so that u_{n-1} is a consequence of u_n for all $n = 2, 3, \ldots$. Following Kleiman, we shall construct a series of groups C_m, $m = 1, 2, \ldots$, such that every u_n holds in some C_m, with $m = m(n)$ depending on n, while none of the groups C_m lies in $\mathfrak{B}_4\mathfrak{B}_2$; that is, each group C_m fails to satisfy some u_n, $n = n(m)$. It is then clear that the system u_n, $n = 1, 2, \ldots$, is not equivalent to any finite subsystem, that is, to the identity in the subsystem with greatest numeral. Elementary considerations show that this means precisely that there is no finite basis for the identities. The group C_m is constructed as a semidirect product of groups A_m and B_m (see 2.1), where $B_m \lhd C_m$. For A_m we take the group

$$A_m = \langle a_1, \ldots, a_{2m}; a_i^2 = 1, [[a_i, a_j], a_k] = 1, i, j, k = 1, \ldots, 2m \rangle.$$

It is not hard to check that A_m is nilpotent of class 2 (see § 4), $A_m^2 = [A_m, A_m]$ and $[A_m, A_m]$ is the direct product of the cyclic subgroups generated by the $[a_i, a_j]$ with $i > j$; that is, $|[A_m, A_m]| = 2^{\binom{2m}{2}} = 2^{m(2m-1)}$.

We denote the image of the element g of A_m under the homomorphism $A_m \to A_m/[A_m, A_m]$ by $\bar{g}$. Choose an element a of $[A_m, A_m]$, and set $\rho(a) = 1$, $\rho(g) = 0$ when g is any element of $[A_m, A_m] \setminus \{a\}$.

The group B_m is generated by elements b^g, c^k, where $g \in A_m$, $k \in A_m/[A_m, A_m]$, with defining relations $[c^k, b^g] = (b^g)^2 = (c^k)^2 = 1$ for all g and k, $[b^g, b^h] = 1$ whenever $\bar{g} \neq \bar{h}$. When $\bar{g} = \bar{h}$, $[b^g, b^h] = c^{\rho(gh^{-1})}$. The action of A_m on B_n is as follows: if $g_1 \in A_m$, then $g_1 b^g g_1^{-1} = b^{g_1 g}$, $g_1 c^k g_1^{-1} = c^{\bar{g}_k}$. It is not hard to check that this is a good definition of a homomorphism $A_m \to \operatorname{Aut} B_m$. We get also that $D_m = [A_m, A_m]B_m \subseteq C_m$. Every element d of D_m has an expression $d = gb^{g_1} \dots b^{g_n}$, where $g \in [A_m, A_m]$ and $g_1, \dots, g_n \in A_m$. We claim that $d^4 = 1$ if and only if either $\rho(g) = 0$, that is, $g \neq d$, or the number of g_i occurring in the expression for d such that $\bar{g}_i = k \in A_m/[A_m, A_m]$ is even. It can be assumed that all the $\bar{g}_i$ are equal to k, since b^{g_i} and b^{g_j} commute if $\bar{g}_i \neq \bar{g}_j$. Since $uv = [u, v]vu$ and commutators in B_m are central and of order 2, the elements of B_m can be rearranged after squaring. We have $d^2 = b^{gg_1} \dots b^{gg_n} b^{g_1} \dots b^{g_n}$ and $d^4 = (b^{g_1} b^{gg_1} \dots b^{g_n} b^{gg_n})(b^{g_1} b^{gg_1} \dots b^{g_n} b^{gg_n})$. The pairs b^{g_i}, b^{gg_i} commute, as can be checked by a simple calculation, while for their squares we have $(b^g i b^{gg} i)^2 = [b^{g_i}, b^{gg_i}] = (c^k)^{\rho(g)}$. Therefore, we get that $d^4 = (c^k)^{n\rho(g)}$, and since c^k has order 2, this proves everything.

We consider next the values of the word u_n in C_m: $u_n = ((g_1 b_1)^2 \dots (g_n b_n)^2)^4 = (g_1^2 \dots g_n^2 \tilde{b})^4$. This value is identically 1 if and only if a cannot be written in the form $a = g_1^2 \dots g_n^2$ for any $g_1, \dots, g_n$ in A_m. For, by what has gone before, if a cannot be written in that form, we have $\rho(g_1^2 \dots g_n^2) = 0$ and $u_n \equiv 1$ identically in C_m. If $g_1^2 \dots g_n^2 = a$, we have $u_n(g_1, \dots, g_{n-1}, g_n b^1) = (ab^{g_n} b^1)^4 \neq 1$ (we can assume that $g_n \notin [A_m, A_m]$). Finally, there exists an m such that some element of $[A_m, A_m]$ cannot be written as a product of n squares. For example, this is so when $m = n + 1$. Indeed, the order of $[A_m, A_m]$ is $2^{m(2m-1)}$. Since the elements of A_m can be rearranged on squaring, g^2 is associated with the set of those elements a_i that occur in g an odd number of times, that is, there are no more elements g^2 than there are subsets of $\{a_1, \dots, a_{2m}\}$, that is, 2^{2m}. Finally, there are not more than $2^{2mn} = 2^{2m(m-1)} = 2^{2m^2-2m}$ elements of the form $g_1^2 \dots g_n^2$, that is, less than the order of $[A_m, A_m]$. Consequently, for each n there is a $C_m = C_m(a)$ in which $u_n \equiv 1$ identically.

Note that this variety is insoluble, since the variety $\mathfrak{B}_4$ of groups of exponent 4 is insoluble (see § 5). However, all the C_m are soluble of length 4, so that the identities $u_n \equiv 1$ are not consequences of finitely many of them in the variety of soluble groups of length 4.

There is a combinatorial device due to Higman, first applied by Cohen to metabelian groups. The result is:

Theorem. *Every metabelian group has a finite basis for its identities.*

Bryant and Newman developed this method and found a positive solution to the finite basis problem for the variety of groups whose commutator subgroups are nilpotent of class 2; Vaughan-Lee did the same things for groups that are simultaneously nilpotent-by-abelian and abelian-by-nilpotent; finally, Krasil'nikov and Shmel'kin for groups that are extensions of nilpotent groups by abelian groups of finite exponent. However, it is believed that the possibilities of the method are not yet exhausted, and that it should be within reach to find a basis for the identities of any group with nilpotent commutator subgroup; more precisely, to get the analogue of the result proved by Krasil'nikov for Lie algebras over a field of characteristic zero. The problem is also still unsolved for matrix groups: we recall that, by the Tits alternative and the Kolchin-Mal'tsev theorem (see 1.1.4 and 2.3.2), every matrix group satisfying a nontrivial identity has a subgroup of finite index with nilpotent commutator subgroup.

It is clear from what has been said that the variety of metabelian groups has certain advantages over other varieties. However, the attempts to describe all its subvarieties has not yet met with success (Kovács, Bryce): there is a calculation of the subvarieties whose free groups are torsionfree, but the problem is completely unintelligible for p-groups.

As for general varieties of soluble groups, as a rule they contain all metablelian groups. This result is due to Kargapolov and Churkin, and independently to Groves:

Theorem. *A variety of soluble groups either contains all metabelian groups, or else every group G in it has a normal series $G \rhd H \rhd K$ for which G/H and K are of finite exponent and H/K is nilpotent.*

Groves has described other soluble varieties on the principle of "what they do not contain" (a dichotomy principle suggested by Phillip Hall).

Between metabelian groups and more general soluble groups, the *centre-by-metabelian* groups (see 3.4) occupy an important position. This property is defined by the identity $[[[x, y], [z, u]], v] = 1$. A very unexpected result was the discovery by Kanta Gupta of torsion in the free centre-by-metabelian group G_n of free rank $n \geqslant 4$: it contains an elementary abelian 2-group of order $2^{\binom{n}{4}}$.

Starting from the observation that the second derived group G_n'' of G_n can be obtained from the $\mathbb{Z}[B]$-module G_n'/G_n'' (where $B = G_n/G_n'$) using an operation similar to the exterior product over the group algebra $\mathbb{Z}[B]$, Kuz'min used the apparatus of homological algebra to study G_n, and later groups of a more general form. Let $\mathfrak{V}_l$ be the variety of groups in which the central factor-groups are soluble of length at most l, and $G_n(l)$ the free group of rank n of this variety. Then, for $l > 2$ and $n > 1$, the periodic part of $G_n(l)$ is an elementary abelian 2-group of countably infinite rank. Kuz'min's methods apply to the universal central extension of the group F/N' (see 1.4.3), that is, to groups of the form $F/[N', F]$. It turns out that 2 and 4 are the only possibilities for the orders of periodic elements. If F/N has no elements of order 2, the periodic part of $F/[N', F]$ is isomorphic to $H_4(F/N, \mathbb{Z}_2)$, that is, in this case the periodic part is of exponent 2 if it is not trivial. This last result can be viewed as a description of the periodic part of the Schur multiplier of F/N', that is, of $H_2(F/N')$. The groups $H_n(F/N')$, $n > 2$, also have fairly detailed descriptions.

Theorem. *If p and q are primes and the factor-group F/N has no elements of order p, the q-component of $H_p(F/N')$ is isomorphic to $H_{p+2}(F/N, \mathbb{Z}_p)$ for $q = p$, and is zero when $q \neq p$, 2.*

The technique in the proof of this theorem can be used to reveal other deep-lying facts about the homology groups of groups of the form F/N'.

The reader can become acquainted with other questions in the theory of varieties by consulting H. Neumann (1967) or the survey Bakhturin and Ol'shanskij (1988) in this series. It is typical that important results are often obtained at the junction of combinatorial and structural methods.

The fundamental algorithmic problems posed for the class of all groups (see 1.4.1) also make sense in the class of all soluble groups. Sometimes they are more interesting in this context. While they usually have negative answers in the class of all groups, they could have positive solutions in the class of soluble groups, and then it is possible that there are differences according to the soluility length or in which varieties the groups happen to lie. It is a remarkable fact that there is great activity in this part of structural group theory.

We note first that if G has a recursively enumerable system of generators and defining relations (see 1.4.1), all consequences of the relations can be effectively written as an infinite sequence. Thus, if some equation is really a consequence, sooner or later we will be convinced of it. The set of all finite groups K_i can be effectively written in order, and if the given group is residually finite (and there is an algorithm for deciding if a finite group is a homomorphic image of G: this is always the case when G is finitely related), every nonidentity word in G will be detected in the sequence K_i. The simultaneous prosecution of these two procedures is the basis for a positive solution of the word problem. (We have reproduced here an important observation of Mal'tsev, one which emphasises the value of residual finiteness.) The algorithmic word problem has the positive solution for all the residually finite groups (such as polycyclic and metabelian groups) considered earlier. A very nontrivial problem is the isomorphism problem for nilpotent groups. The solution was found by Grunewald and Segal, and is based on the existence of certain algorithms in arithmetic matrix groups.

The first successful step in this area was taken by Remeslennikov, who constructed a group that is finitely presented in the variety of soluble groups of length 5 (that is, it is obtained by imposing finitely many further relations on a free soluble group of length 5) with insoluble word problem. Later, Kukin proved that the word problem is insoluble in any variety containing the product variety $\mathfrak{N}_3\mathfrak{A}$ (that is, the variety consisting of extensions of nilpotent groups of class at most 3 by abelian groups), while Kharlampovich showed that it is soluble in $\mathfrak{N}_2\mathfrak{A}$ and all its subvarieties. Harder still was the problem of constructing a finitely presented group (finitely presented in the class of all groups, that is, in the sense of §4 of Chapter 1) satisfying a nontrivial identity. Kharlampovich solved this question.

Theorem. *There exists a finitely presented soluble group of length* 3 *with nilpotent commutator subgroup and insoluble word problem.*

In constructing her example, Kharlampovich used a method of interpretation in a species of Turing machine (a two-tape Minsky machine computing any given partial recursive function) in the group and codifying elements of the group using an auxiliary semigroup.

Noskov showed that the elementary theory $T(G)$ of a finitely generated soluble group is decidable (that is, there exists an algorithm for determining the truth of a closed formula in restricted predicate calculus involving the group operation) if and only if G has an abelian subgroup of finite index. This theorem solves a problem of Kargapolov and completes a cycle of investigation initiated by Ershov and Romanovskij. The essential thing here is a proof that the elementary theory $T(G)$ of a group G having no abelian subgroup of finite index is undecidable. There is a reduction to the metabelian case, indeed to a group H with a formal abelian normal subgroup M with abelian factor-group H/M such that the annihilator L of the module M in the ring $\mathbb{Z}[H/M]$ is a simple ideal and the $\mathbb{Z}[H/M]/L$-module M is torsionfree. (A subgroup is said to be *formal* if it consists of precisely those elements that satisfy some formula in restricted predicate calculus in a single free variable – for example, the centre of group is defined by the formula $\forall x(xy = yx)$.) After this reduction, the arithmetic of the natural numbers can be interpreted in G: and it is known that it has undecidable theory.

The term "interpret" can be explained by a simpler example. If R is a (not necessarily associative) ring with identity, it has associated with it a group consisting of triples of elements with multiplication

$$(a, b, c)(x, y, z) = (a + x, b + y, bx + c + z).$$

This group is nilpotent of class 2 (see the following section) and the fact that a formula is true in it means that some formula is true in R. Conversely, let G be a free nilpotent group of class 2, that is, the factor-group $F(a, b)/\gamma_3(F(a, b))$ of a free group by the third term of its lower central series. Then the commutator subgroup of G can be converted into the ring of integers (for fixed a, b); addition is just group multiplication, and multiplication is give by the rule $z \times w = [x, y]$, where $[x, b] = 1, [y, a] = 1, z = [a, x], w = [b, y]$. The fact that commutation in groups of class 2 is distributive (see the following section) over multiplication yields the distributive law in the ring. It is easy to see from the formula just indicated that z will correspond to the integer m such that $z = [a, b]^m$, and w with the integer n such that $z \times w$ corresponds to mn. The identity element is $[a, b]$.

When $R = \mathbb{Z}$, we get that the elementary theory of the group G is identical with that of the ring $\mathbb{Z}$, which is undecidable; that is, G has undecidable theory. This argument of Mal'tsev was the starting-point for all subsequent results.

It is by no means the case that the free groups of all soluble varieties are residually finite (Kleiman). Even more, Kleiman has constructed a soluble variety such that there is no algorithm for recognizing equalities in its free groups, thus providing the answer to an open problem.

Theorem. *There is no algorithm for recognizing the consequences of a given identity.*

This is the final nail in the coffin of the idea that there may be a classification of arbitrary identities.

§ 4. Nilpotent Groups

4.1. General Properties. The very name indicates the connection between groups of this type and rings. In fact, formal adjunction of an identity to any nilpotent associative ring A (that is, a ring in which there is an integer n such that $a_1 a_2 \ldots a_n = 0$ for all $a_1, a_2, \ldots, a_n$ in A) gives the nilpotent group consisting of all elements of the form $1 + a$, $a \in A$. There is an even closer connection between nilpotent groups and nilpotent Lie algebras – right up to the automatic restatement of properties proved for the objects in certain subclasses. The nilpotency of the Sylow subgroups of finite groups adds many important examples and testifies to the overall importance of this concept. Nilpotent groups occupy a vital place in group theory, even though they are intermediate between abelian and soluble groups – in definition as in essence.

Definitions.

1) The *lower central series*

$$G = \gamma_1(G) \supseteq \gamma_2(G) \supseteq \cdots \tag{1}$$

of a group is the normal series in which $\gamma_{n+1}(G) = [\gamma_n(G), G] = \langle [a, b] | a \in \gamma_n(G), b \in G \rangle$ for $n \geqslant 1$.

2) A group G is said to be nilpotent if $\gamma_{n+1}(G)$ is the identity subgroup for some n. If in addition $\gamma_n(G) \neq 1$, G is said to be nilpotent of class n.

As is clear from the definition, nilpotency is stronger than solubility: the factor-group $\gamma_i(G)/\gamma_{i+1}(G)$ is not merely abelian, it is in the centre of $G/\gamma_{i+1}(G)$. More generally, a normal series $G = G_1 \supseteq G_2 \supseteq \cdots$ is said to be a *central series* if G_i/G_{i+1} is in the centre of G/G_{i+1} for each i. It is easily checked by induction that $\gamma_i(G) \subseteq G_i$ whenever $G = G_1 \supseteq G_2 \supseteq \cdots$ is a central series – this explains the word "lower" in the definition of the series (1). The class of all nilpotent groups of class at most n can be given by the single identity $[\ldots[[x_1, x_2], x_3], \ldots, x_{n+1}] = 1$.

It is an important fact that k-fold commutation in groups of class k is distributive over multiplication. We prove this for $k = 2$. We have the identity $[x_1 x_2, y] = x_1 x_2 y x_2^{-1} x_1^{-1} y^{-1} = x_1 [x_2, y] x_1^{-1} [x_1, y]$: since commutators are central in groups of class 2, the right-hand side is $[x_1, y][x_2, y]$. This gives distributivity in the first argument; distributivity in the second argument is proved in the same way.

This property explains why in a group G on generators a_i, $i \in I$, the k-th term of the lower central series is the normal closure of the set of k-fold commutators $[\ldots[a_{i_1}, a_{i_2}], \ldots]$ in the generators with consecutive distribution of brackets (such commutators are said to be left-normed). For the proof, it is enough to go over to the group $G/\gamma_{k+1}(G)$ and apply the distributive law. We also see that, for finite I, the central factors $\gamma_k(G)/\gamma_{k+1}(G)$ are finitely generated, so that finitely generated nilpotent groups are polycyclic (see 3.2). This property lies at the heart of a lemma of Burnside: if G is nilpotent, then a set of elements generates

G if it does so together with the commutator subgroup $[G, G]$. It also explains the fact that the product of p-elements in a nilpotent group is a p-element. Let us prove this for $k = 2$. If $a^{p^\alpha} = b^{p^\beta} = 1$ and $\alpha \geqslant \beta$, then $(ab)^{p^\alpha}$ lies in $[G, G] = \gamma_2(G)$. The commutator subgroup of $\langle a, b \rangle$ is generated by $[a, b]$, so that $(ab)^{p^\alpha} = [a, b]^s$ for some s. But $[a, b]^{p^\alpha} = [a^{p^\alpha}, b] = 1$ by distributivity, so that ab is a p-element. We have, therefore:

Theorem. *The periodic part of a nilpotent group is a subgroup which is the direct product of its primary components.*

As with abelian groups, this result reduces the study of nilpotent groups, in some sense, to the study of torsionfree groups and p-groups.

Every finite p-group G (a group of order p^k, p prime) is nilpotent. We prove this now, noting first that the number of elements conjugate to a given element a is the index of the centralizer $C(a) = \{g \in G | ga = ag\}$. In fact $g_1 a g_1^{-1} = g_2 a g_2^{-1} \Leftrightarrow g_2^{-1} g_1 a g_1^{-1} g_2 = a \Leftrightarrow g_2^{-1} g_1 \in C(a) \Leftrightarrow g_2^{-1} g_1 C(a) = C(a) \Leftrightarrow g_1 C(a) = g_2 C(a)$. This establishes a bijection between the cosets modulo the centraliser and the elements conjugate to a. Therefore, every nontrivial finite p-group has nontrivial centre, since otherwise the decomposition of G into its conjugacy classes, $G = \{1\} \cup Cl(a_1) \cup \cdots \cup Cl(a_t)$, yields that $|G| = 1 + p^{\alpha_1} + \cdots + p^{\alpha_t}$, where all the α_i are nonzero, and the contradiction is that $p | 1$.

Next, for any group G we construct the *upper central series* by induction: $Z_1(G) = Z(G)$ is the centre, and $Z_{i+1}(G)/Z_i(G) = Z(G/Z_i(G))$. The upper central series grows at least as fast as any other central series in a nilpotent group, and so its length is equal to the nilpotency class.

Returning now to the case of a finite p-group G, we see that the nontriviality of the centre enables us to conclude by induction that the upper central series reaches the whole group: $Z_s(G) = G$, so that the group is nilpotent.

Of the general properties, we single out the following results.

Theorem (P. Hall). *If N is a normal nilpotent subgroup of a group G such that the factor-group $G/[N, N]$ of G by the commutator subgroup of N is nilpotent, then G itself is nilpotent.*

Theorem (Fitting). *If H and K are nilpotent normal subgroups of a group, then the product HK is nilpotent, of class not more than the sum of the classes of H and K.*

4.2. Torsionfree Nilpotent Groups. A good example of a torsionfree nilpotent group is the group $UT_n(k)$ of all upper-triangular matrices with 1 down the main diagonal, where k is $\mathbb{Z}$ or any other ring without additive torsion. We establish the absence of torsion in $G = UT_n(Z)$ as an example. Every matrix in G can be written in the form $E + A$, where E is the identity matrix and A is a nilpotent matrix (it has zeros down the main diagonal). We have

$$(E + A)^k = E + kA + \frac{k(k-1)}{2!}A^2 + \cdots + A^k.$$

When A is raised to a power, the nonzero entry nearest the main diagonal gets further away from it (the same conclusion applies to the multiplication of different nil-triangular matrices). Thus, if a_{ij} is the nearest nonzero entry to the diagonal, the entry at that spot in $(E + A)^k$ is ka_{ij}; and this means that $(E + A)^k \neq E$ if $A \neq 0$.

A central series for G is formed by the subgroups $UT_n^m(\mathbb{Z})$ consisting of the matrices (a_{ij}) such that $a_{ij} = 0$ for $i < j < i + m$ $(UT_n^1(\mathbb{Z}) = G)$. Because, if $E + A \in UT_n^m(\mathbb{Z})$ and $E + B \in UT_n^l(\mathbb{Z})$, then $[E + A, E + B] = E + AB - BA + \cdots$, where the dots stand for products of matrices in which there are more than two factors, so that the commutator lies in $UT_n^{m+l}(\mathbb{Z})$.

This example is a very important one, as is shown by the following result of Phillip Hall.

Theorem. *Every finitely generated torsionfree nilpotent group can be embedded in $UT_n(\mathbb{Z})$ for suitable n.*

The group $UT_n(k)$ can be viewed as the set of all linear operators inducing the identity on the factors of the series

$$\{0\} \subseteq \langle e_1 \rangle \subseteq \langle e_1, e_2 \rangle \subseteq \cdots \subseteq k^n,$$

where $e_1, \ldots, e_n$ is a basis of k^n. Kaluzhnin obtained some considerable generalisations of this property.

Theorem. *Let G be the group of all automorphisms fixing some normal series*

$$\{1\} = A_0 \subseteq A_1 \subseteq \cdots \subseteq A_n = A$$

of subgroups of an arbitrary (nonabelian) group A and inducing the identity automorphism on all the factors A_i/A_{i-1}. Then, G is nilpotent.

Hall removed the normality condition from this statement and gave $n(n - 1)/2$ as an upper bound for the nilpotency class.

Torsionfree nilpotent groups have the property that extraction of roots is unique in them (if $x^n = y^n$, then $x = y$). Of course, this property can happen only in torsionfree groups, but not in all of them. For example, in the group $\langle a, b; a^2 = b^2 \rangle$ (which is the generalized free product of two infinite cyclic groups with amalgamation and therefore torsionfree), we have $a \neq b$. The unique root-extraction property enables us to bring in the concept of isolated subgroup.

Definition. A subgroup H of G is said to be isolated if $x \in H$ whenever $x^n \in H$ for x in G and $n > 0$. The isolator $I(A)$ of A is the smallest isolated subgroup of G containing A.

If H is normal in G, then H is isolated if and only if G/H is torsionfree.

In a group with unique extraction of roots, the isolator of a subgroup of the centre is itself contained in the centre. For, if $A \subseteq Z(G)$ and $x^n \in A$, then $yx^ny^{-1} = x^n$ for all y, that is, $(yxy^{-1})^n = x^n$, so that $yxy^{-1} = x$ and $x \in Z(G)$. Together with an inductive argument, this shows that every finitely generated torsionfree nilpotent group has a central series

$$G = A_1 \supseteq A_2 \supseteq \cdots \supseteq A_r \supseteq \{1\}$$

with infinite cyclic factors. If A_i/A_{i+1} is generated by the coset containing a_i, every element of G has a unique expression of the form $g = a_1^{n_1} \dots a_r^{n_r}$, with the $n_i \in \mathbb{Z}$. Thus, we can associate with each element g its "coordinates" $(n_1, \dots, n_r)$, and to construct the group one simply has to compute the coordinates of a product in terms of the coordinates of the factors. It turns out that the coordinates of a product can be expressed as polynomials in the coordinates of the factors. Moreover, the coordinates of g^m are polynomials in $m, n_1, \dots, n_r$. This enables us to suppose that the coordinates come not from $\mathbb{Z}$ but from some larger ring, for example $\mathbb{Q}$, $\mathbb{R}$ or the ring of p-adic integers, and then raise to powers that are elements of this larger ring (see the lectures of P. Hall (1957) for more details). For example, if $\mathbb{R}$ is taken as the base ring, we obtain an embedding of G into a r-dimensional Lie group; while with $\mathbb{Q}$ we get the Mal'tsev completion of the original torsionfree nilpotent group (the definition is just as it was for abelian groups in 1.2).

Theorem (Mal'tsev). *Completions exist for an arbitrary torisonfree nilpotent group G, and any two completions are isomorphic over G.*

This theorem has been rederived many times, and it can be explained from the starting-point that every torsionfree nilpotent group is residually a finite p-group, for each prime p. If $g_1, g_2, \dots$ are all the non-identity elements of such a group G, and $p_1, p_2, \dots$ are all the prime numbers, we shall denote by G_i a finite p_i-group such that there is an epimorphism ϕ_i: $G \to G_i$ taking g_i to a non-identity element. Thus, G is embedded in the Cartesian product P of the G_i, which is nilpotent of the same class as G since the ϕ_i cannot increase the nilpotency class. The factor-group of this Cartesian product by its periodic part (namely, the direct product of the G_i) is divisible and contains G (see 1.5). The remainder of the proof is a matter of not particularly tricky technicalities.

A completion of $UT_n(\mathbb{Z})$ is $UT_n(\mathbb{Q})$. It is divisible since, by the binomial theorem,

$$(E + A)^\lambda = E + \lambda A + \frac{\lambda(\lambda - 1)}{2!} A^2 + \cdots + \frac{\lambda(\lambda - 1)\dots(\lambda - n + 2)}{(n - 1)!} A^{n-1}$$

for each rational λ, since $A^n = 0$, and a suitable integer k can be found such that $(E + A)^k$ is an integer matrix. Hall axiomatised the formation of powers with exponents from an arbitrary binomial ring (that is, one in which the binomial coefficients make sense, for instance $\mathbb{Z}$ and all fields of characteristic zero): 1) $x^1 = x$, 2) $x^{\lambda+\mu} = x^\lambda x^\mu$, 3) $x^{\lambda\mu} = (x^\lambda)^\mu$, 4) $y^{-1}x^\lambda y = (y^{-1}xy)^\lambda$, 5) $x_1^\lambda x_2^\lambda \dots x_n^\lambda = t_1^\lambda t_2^{\binom{\lambda}{2}} \dots t_k^{\binom{\lambda}{k}}$, where k is the nilpotency class of the group $\langle x_1, \dots, x_n \rangle$ and $t_i = \tau_i(x_1, \dots, x_n)$ are the values of certain fixed words associated with the collecting process in nilpotent groups. (For instance, $\tau_1(x_1, x_2) = x_1 x_2$, $\tau_2(x_1, x_2) = [x_1, x_2]$.) In this way, there arises a theory of "modules" whose additive groups are not abelian but locally nilpotent torsionfree groups.

A *D-group* is a group in which every equation $x^n = g$ has a unique solution, where $n \in \mathbb{Z}$, $n \neq 0$. The structure of nilpotent D-groups is somewhat simpler than that of general torsionfree nilpotent groups (though not as simple as for abelian groups – see 1.2). Mal'tsev remarked that the category of such groups is equivalent to the category of nilpotent Lie algebras over $\mathbb{Q}$. If L is a Lie algebra of that sort, the Campbell-Baker-Hausdorff series

$$x \circ y = \ln(e^x e^y) = x + y + \tfrac{1}{2}[x, y] + \tfrac{1}{12}[[y, x], x] + \tfrac{1}{12}[[x, y], y] + \cdots$$

(see Jacobson (1961), for example) has only finitely many nonzero terms, and the operation $\circ$ gives L the structure of a nilpotent D-group (clearly, raising x to the λ-th power for rational λ produces λx). In this set-up the concepts of subalgebra and isolated subgroup coincide, as do those of ideal and isolated normal subgroup. Moreover, every nilpotent D-group can be obtained in this way.

Myasnikov and Remeslennikov have studied the elementary theory of nilpotent D-groups (the collection of formulae in first order predicate calculus that hold in a given group). It is easy to see that a group of this sort can be written in the form $G = G_0 \times G_1$, where G_1 is abelian and the centre of G_0 is contained in its commutator subgroup. Further, two groups G and G' have the same theory if and only if G_0 and G'_0 are isomorphic. Thus, essentially all properties of nilpotent D-groups can be expressed in the language of first-order predicate calculus.

The successes of the theory of nilpotent D-groups has induced several authors to look at other classes of D-groups. Baumslag has determined the structure of free D-groups (in terms of amalgamated free products). The following problem in the theory of D-groups is still open: is every free D-group residually a nilpotent D-group? If the answer is yes, the solution could be just like that for free groups (see 1.1.4), namely the free D-group could be represented in the ring $A(\mathbb{Q}, X)$ of power series with unit free terms – this group is itself a D-group.

There is another central series in a group G, one that can be defined using the group algebra $A[G]$ of G over a commutative ring A with 1: $D_n(A, G) = (I^n + 1) \cap G$, where I is the augmentation ideal (see 1.4.3) of $A[G]$. Since $G/D_n(A, G)$ is a subgroup of the multiplicative group of the nilpotent algebra I/I^n with an externally adjoined identity element, it is a group of class at most $n - 1$, so that $D_n(A, G) \supseteq \gamma_n(G)$. The subgroup $D_n(A, G)$ is called the n-th *dimension subgroup* of G over A. When A is a field of characteristic zero, $D_n(A, G)$ is the isolator of $\gamma_n(G)$, so that $D_n(\mathbb{Z}, G)$ lies between the isolator of $\gamma_n(G)$ and $\gamma_n(G)$ itself. The conjecture that $D_n(\mathbb{Z}, G) = \gamma_n(G)$ in all cases was refuted by Rips in 1972. He constructed an example of a finite 2-group G for which $D_4(\mathbb{Z}, G) \neq \gamma_4(G)$. Sjogren has obtained the following elegant result: for every group G, the exponent of $D_n(\mathbb{Z}, G)/\gamma_n(G)$ divides $b_1^{\binom{n-2}{1}} \ldots b_{n-2}^{\binom{n-2}{n-2}}$, where b_k is the least common multiple of the numbers $1, 2, \ldots, k$. As a corollary, we get the following result. If G is a p-group, then $D_n(\mathbb{Z}, G) = \gamma_n(G)$ whenever $n \leqslant p + 1$. If A is a field of characteristic $p > 0$, then $D_n(A, G) = \prod_{ip^j \geqslant n} \gamma_i(G)^{p^j}$.

4.3. Basic Commutators and Connections with Lie Algebras. We have already observed (in section 4 of Chapter 1) that an element of the free group $F(X)$ lies in $\gamma_n(F(X))$ if and only if the series representing it in $A(\mathbb{Z}, X)$ is of the form $1 + u_n + \cdots$. The absence of torsion in the free nilpotent group $F(X)/\gamma_{c+1}(F(X))$ is a consequence of this fact. We shall now exhibit an explicit basis for each factor of the lower central series, in terms of the so-called basic commutators.

Basic commutators of weight 1 are the free generators, that is, elements of X. We order these in any way at all, and suppose that basic commutators of weight less than n are already defined.

Definition. By a basic commutator of weight n we mean a commutator $[w_1, w_2]$, where w_1 and w_2 are basic commutators of weights n_1 and n_2 such that $n_1 + n_2 = n$, and the following conditions are satisfied: 1) $w_1 > w_2$; 2) if $w_1 = [v_1, v_2]$, then $v_2 \leqslant w_2$. Basic commutators of weight n are assumed to come later than those of weight less than n; the ordering between them is arbitrary.

Thus if $X = \{x, y\}$, the basic commutators of weights up to 3 (assuming that $x > y$) are x, y, $[x, y]$, $[[x, y], y]$, $[[x, y], x]$; but not $[x, [x, y]]$.

The basic commutators of weight n lie in $\gamma_n(F)$, and their cosets form a basis for the free abelian group $\gamma_n(F)/\gamma_{n+1}(F)$. The number of basic commutators of weight n, that is, the rank of this group, is given by the Witt formula – see 1.1.4. As a result, every element g of a free nilpotent group has a unique expression as a product of powers of basic commutators in increasing order: $g = c_1^n \dots c_r^n$ (that is, the row $(n_1, \dots, n_r)$ gives the coordinates of g in the sense of 4.2). This representation holds in an absolutely free group only modulo a term of the lower central series.

As well as the definition just given, which is due to M. Hall Jr., there are other definitions of basic commutators (for example, that due to Shirshov). They may differ from each other, but they all have the same significance: they constitute bases for the factors of the lower central series of free groups. Moreover, they are usually obtained *via* Lie algebras. Magnus and Witt proved that the factors of the lower central series of a free group are isomorphic to those of the free Lie ring on the same generators. This has prompted investigators to establish a similar connection between free groups and rings in certain varieties.

If

$$G = A_1 \supseteq A_2 \supseteq \cdots \supseteq A_k \supseteq \cdots \tag{2}$$

is a central series such that $[A_i, A_j] \leqslant A_{i+j}$ in all cases (for example, the lower central series, or the series consisting of the isolators of the terms of the lower central series), we can associate with the series (2) for G a Lie ring whose additive group is the direct sum $\sum_{i=1}^{\infty} A_i/A_{i+1}$ of the factors, and the Lie product is defined as follows. If $g \in A_i/A_{i+1}$ and $h \in A_j/A_{j+1}$ and $\bar{g}$, $\bar{h}$ are inverse images of them in G, then $[\bar{g}, \bar{h}] \in A_{i+j}$ and the coset $[\bar{g}, \bar{h}]A_{i+j+1}$ is the product of g and h in the ring. The multiplication is then extended by distributivity to the whole of the ring. The verification that we have indeed got a (graded) Lie ring is not totally obvious – proof of the Jacobi identity in the Lie ring needs the Witt identity in groups.

Of course, nothing follows from the mere fact of associating a Lie ring with a group; some additional conditions are always necessary. However, it is important firstly from a methodological standpoint, and secondly because it facilitates calculations – additional commutators that have to be considered in the group case, or at the very least written out, simply disappear.

If the group is residually torsionfree nilpotent, its isolated lower central series intersects in the identity, and the associated Lie ring has no additive torsion. Tensoring it with $\mathbb{Q}$ gives a Lie algebra over $\mathbb{Q}$. If we begin with the free group of some variety, we will end up with the free algebra of some variety, which can then be associated with the group variety. This correspondence is injective on the varieties whose free groups are residually torsionfree nilpotent (for example, the variety of all soluble groups of length at most l corresponds to the variety of all soluble Lie algebras of length at most l). However, it is not surjective: no group with a nontrivial identity can give rise to a free algebra of a variety containing $\mathrm{sl}_2(\mathbb{Q})$.

The correspondence just introduced can be used to discover new classes of residually torsionfree nilpotent groups. For example, the set of varieties whose free groups have this property is closed under products. (The product $\mathfrak{U}\mathfrak{V}$ of two varieties $\mathfrak{U}$ and $\mathfrak{V}$ is the class of all extensions of groups in $\mathfrak{U}$ by groups in $\mathfrak{V}$.) Also closed under multiplication is the set of varieties whose free groups have lower central factors that are torsionfree – free abelian – that is, for which a construction of "basic commutators" is possible. Thus, in the free metabelian group $F(X)/F(X)''$, the leftnormed basic commutators of weight k comprise a basis of the k-th lower central factor. This can be generalised to free soluble and indeed to more general groups.

We can consider the Lie algebra given by the above recipe from the series

$$D_1(\mathbb{Z}_p, G) \subseteq D_2(\mathbb{Z}_p, G) \supseteq \cdots .$$

In this case, it can be viewed as a p-algebra (see the relevant volume in this series or Jacobson (1961)), if the p-map is defined as follows. If $g \in D_i(\mathbb{Z}_p, G)$, then $g^p \in D_{p^i}(\mathbb{Z}_p, G)$, so that for $g \in D_i(\mathbb{Z}_p, G)\setminus D_{i+1}(\mathbb{Z}_p, G)$, we define $(gD_{i+1}(\mathbb{Z}_p, G))^{[p]}$ to be $g^p D_{p^i+1}(\mathbb{Z}_p, G)$. That is how the p-map is defined on "homogeneous" elements; it is extended to sums using properties of p-maps.

4.4. Generalised Nilpotency. There are several generalisations of the concept of nilpotent group. However, the ones that have turned out to be the most enduring are local nilpotency and the Engel property.

Definition. A group is said to be locally nilpotent if all of its finitely generated subgroups are nilpotent.

Plotkin obtained the following generalisation of Fitting's theorem (see 4.1).

Theorem. *The product of any two locally nilpotent normal subgroups of a group G is locally nilpotent.*

This makes it possible to distinguish the locally nilpotent radical – the largest locally nilpotent normal subgroup.

Every nilpotent group satisfies the *normaliser condition*: if H is a proper subgroup of G, the normaliser $N(H)$ of H in G is strictly larger than H. It is easy to establish by induction on the length of a central series (for finite groups, the normaliser condition is equivalent to nilpotency). More can be said: every subgroup of a nilpotent group is part of a subnormal series (see (1) in § 3). An example of Heineken and Mohamed shows that the converse is false – there exists a (metabelian) non-nilpotent group in which every subgroup is subnormal. However, we have:

Theorem (Plotkin). *Every group satisfying the normaliser condition is locally nilpotent.*

On substituting $x_2 = y, \dots, x_{n+1} = y$ in the identity $[x_1, x_2, \dots, x_{n+1}] = 1$ defining nilpotency of class at most n, we get the weaker identity

$$[\dots[[x, y], \dots, y] = 1. \tag{3}$$

Definition. Groups satisfying identity (3) are said to be n-Engel (or simply Engel-) groups.

Identity (3) is a corollary of nilpotency, and at the same time it is the simplest means of attempting to guarantee it (there are just two variables here). A still unsolved problem asks: is every Engel group locally nilpotent?

Every Engel group with $n = 2$ is even nilpotent. Clearly, the identity $[[x, y], y] = 1$ holds if and only if every element y commutes with all its conjugates xyx^{-1}, so that the normal closure of every element is commutative. A little further argumentation shows that the group must be nilpotent of class at most 3.

However, even for $n = 3$ nilpotency does not follow. Consider an infinite abelian group G of exponent 2 and its group algebra $\mathbb{Z}_2[G]$ over the field $\mathbb{Z}_2$ of order 2. An example of the type we want is the group H of matrices of the form $\begin{pmatrix} g & u \\ 0 & 1 \end{pmatrix}$ with $g \in G$, $u \in \mathbb{Z}_2[G]$, as mentioned in 1.4.3. This group is 3-Engel, since if $[x, y] = \begin{pmatrix} 1 & t \\ 0 & 1 \end{pmatrix}$, $y = \begin{pmatrix} g & u \\ 0 & 1 \end{pmatrix}$, we have

$$[x, y, y, y] = \begin{pmatrix} 1 & (g-1)^2 t \\ 0 & 1 \end{pmatrix},$$

while $(g - 1)^2 = g^2 - 2g + 1 = 0$ in $\mathbb{Z}_2[G]$. However, H is not nilpotent, and indeed it is an easy exercise to check that H has trivial centre. Nevertheless, H is locally nilpotent. This is not an accident, since we have:

Theorem (Heineken). *Every* 3-*Engel group is locally nilpotent.*

One can extend the concept of Engel group by allowing the number n in equation (3) to depend on x and y. In this case, a remarkable example due to

Golod (see § 5) shows that such a group need not be locally nilpotent. Moreover, the example shows that for every $d > 1$ there is a non-nilpotent $(d + 1)$-generator group all of whose d-generator subgroups are nilpotent (of varying classes in general). However, in the case of soluble groups, every "unbounded Engel" group is locally nilpotent.

The concept of Engel group arose by analogy with the similar concept for Lie algebras. Up to the present time, the norm is that our knowledge of Engelicity in groups is based on results in Lie algebras. In his study of the associated Lie rings of groups of exponent p, Magnus established that they satisfy the $(p - 1)$-st Engel condition. This opened up an approach for solving the *restricted Burnside problem* for exponent p. For exponent n it asks the question: is there a bound on the orders of the d-generator finite groups satisfying the identity $x^n = 1$, in terms of d and n? In the case of prime exponent p, it is enough to show that a $(p - 1)$-Engel Lie algebra over a field of characteristic p is locally nilpotent.

Theorem (Kostrikin). *Every finitely generated* $(p - 1)$*-Engel Lie algebra over a field of characteristic p is locally nilpotent.*

The restricted Burnside problem for prime exponent p thus has the positive solution.[4] (See the survey Bakhturin and Ol'shanskij (1988) in this series for more details.) It follows that the class of locally nilpotent (or equivalently, locally finite) grups of exponent p form a variety, called in H. Neumann (1967) the Kostrikin variety. The nilpotency problem for these varieties was solved in the negative by Razmyslov in the cases $p \geqslant 5$. Thus, as the number of generators grows, the nilpotency class of a finite group of exponent p can increase without limit. There is a complicated dependency between the Engel degree n of a nilpotent p-group and the numbers p, under which the nilpotency class of an n-Engel group can in fact increase as the number of generators increases. As the example constructed above shows, for $n = 3$, we have $p = 2$, and $p = 5$ as well, as was proved by Bachmuth, Mochizuki and Walkup independently of Razmyslov. For $n = 3$ and other primes p, nilpotency is guaranteed (Heineken).

Zel'manov has shown recently that every n-Engel Lie algebra over a field of characteristic zero is nilpotent (local nilpotency was established by Kostrikin). The same result holds for finite characteristic p when p is large enough in comparison with n. From this and the connection between groups and Lie algebras described above, it follows that every n-Engel torsionfree locally nilpotent group is nilpotent (the same is true for n-Engel p-groups for almost all p).

4.5. Finite Factor-Groups. Every finitely generated nilpotent group is residually finite and finitely presented. As we remarked in 3.6, the word problem is always soluble in such a situation. Solution of the conjugacy problem is bound up in many instances with the following group property.

[4] The restricted Burnside problem for prime-power exponents has recently been solved by E.I. Zel'manov.

Definition. A group G is said to be conjugacy separable if any two of its elements are conjugate in G if and only if their images in every finite homomorphic image of G are conjugate.

This property enables one to solve the conjugacy problem, just as the (weaker) property of residual finiteness does for the word problem (see 3.6).

The first nontrivial result is this direction was discovered by Blackburn.

Theorem. *Every finitely generated nilpotent group is conjugacy separable.*

This result has a development resulting in the following result due to Remeslennikov:

Theorem. *Every polycyclic group is conjugacy separable.*

It could be said that this result is the limit of what is possible in this direction. This is clear from the following example of a finitely generated group (which is metabelian and of finite rank – very nearly polycyclic) constructed by Kargapolov and Timoshenko. Their group G is the subgroup of $GL_3(\mathbb{Q})$ generated by the matrices

$$A = \begin{pmatrix} 1 & 1 & 0 \\ 0 & 1 & 0 \\ 0 & 0 & 1 \end{pmatrix}, \qquad B = \begin{pmatrix} 1 & 0 & 0 \\ 0 & 1 & 1 \\ 0 & 0 & 1 \end{pmatrix}, \qquad C = \begin{pmatrix} p & 0 & 0 \\ 0 & 1 & 0 \\ 0 & 0 & 1 \end{pmatrix}.$$

The elements A and $A[A, B]^{1/p}$ are not conjugate in G, but are conjugate modulo every normal subgroup of finite index. Similar examples have been found by Wehrfritz.

Nevertheless, the algorithmic conjugacy problem is soluble in metabelian groups (Noskov).

We mention here some other classes of conjugacy separable groups, even though they are not formally connected with the conjugacy problem. These are the free products of groups with the property (Remeslennikov, Stieb), the free soluble groups (Remeslennikov and Sokolov), and, more generally, free polynilpotent groups (Kolmakov).

The richness of the structure of the lattice of subgroups of finite index in finitely generated nilpotent groups prompts the thought that residual finiteness can be used to solve the third classical algorithmic problem, namely the isomorphism problem (see 1.4.1). Also, it should be possible to solve it by surveying the finite homomorphic images of the two groups on the one hand, and applying Tietze transformations (see 1.2.1) to presentations for them on the other hand. However, a simple example due to Baumslag reveals the weakness of such a naïve approach: the nilpotent groups

$$G_1 = \langle a, b; a^{25} = 1; bab^{-1} = a^6 \rangle, \qquad G_2 = \langle c, d; c^{25} = 1, dcd^{-1} = c^{11} \rangle$$

are non-isomorphic but have the same stock of finite homomorphic images.

Two groups with identical sets of finite factor-groups are said to be of the same *genus*. (We provide a fairly simple alternative formulation for the reader familiar with the requisite terminology: two finitely generated residually finite groups are of the same genus if and only if their profinite completions are isomorphic.) With the above example in the background, the following powerful result of Pickel is quite unexpected.

Theorem. *Every genus contains only finitely many nonisomorphic finitely generated nilpotent groups.*

A consequence is a positive solution for the isomorphism problem for a given finitely generated nilpotent group G.

The methods of proof of this result forces one to refer to the themes of "linear" or "algebraic" groups – one of the fundamental results being Borel's theorem on the adèle groups of algebraic groups.

Fundamental new difficulties are encountered on proceeding to arbitrary polycyclic groups. Grunewald, Pickel and Segal solved the genus problem for polycyclic groups by combining their results and intensifying their efforts:

Theorem. *The genus of every polycyclic group contains only finitely many isomorphism classes.*

As we remarked earlier, Grunewald and Segal solved the isomorphism problem for nilpotent groups.

§ 5. Periodic Groups

5.1. Statement of the Burnside Problem. We recall the following definitions.

Definitions.

1) A group is said to be periodic if each of its elements has finite order: $(\forall g \in G)(\exists n \in \mathbb{N})(g^n = 1)$.

2) G is said to be of finite exponent if the orders of its elements are bounded: $(\exists n \in \mathbb{N})(\forall g \in G)(g^n = 1)$. We write $G^n = \{1\}$ in this situation.

Let k be a field of characteristic zero, for example, the fields of real or complex numbers.

Theorem (Burnside). *If H is a subgroup of the full matrix group $\mathrm{GL}_m(k)$ of finite exponent (say $H^n = \{1\}$), then H is finite.*

Passing to $\bar{k}$ if necessary, we can assume k to be algebraically closed, and for a proof of this theorem one notes first that the eigenvalues of every matrix h in H are roots of unity, since $h^n = 1$ for all h in H. This means that the traces $\operatorname{tr} h$ of the matrices in H can assume only finitely many different values. Next, an easy induction on the dimension m reduces the problem to the irreducible case, and then general properties representations can be used to select m^2 linearly independent matrices $h(1), \ldots, h(m^2)$ in H. Let h_{ij} denote the (i, j)-entry of any matrix h, $1 \leqslant i, j \leqslant m$. The equations

$$\operatorname{tr}(h(i)h) = \sum_{j=1}^{m} (h(i)h)_{jj} = \sum_{j,k=1}^{m} h(i)_{jk} h_{kj}, \qquad i = 1, \ldots, m^2, \tag{1}$$

can be viewed as a system of m^2 linear equations in the h_{kj}. By the choice of the matrices $h(i)$, the rows of the matrix of this sytem are linearly independent, and it follows that the solution of (1) is unique. The number of systems of this type is finite, since as we have remarked the free terms can have only finitely many values. This means that $|H| < \infty$.

The example of the group $H = UT_2(k)$ of unitriangular matrices over an infinite field k of characteristic $p > 0$ shows that the assumption on the characteristic is essential, since $h^p = 1$ for every element h of H.

For finitely generated groups, the requirement that the group be of finite exponent can be omitted from the statement of Burnside's theorem.

Theorem (Schur). *Every finitely generated periodic subgroup G of $\mathrm{GL}_m(k)$ is finite.*

For the proof, one adjoins to $\mathbb{Q}$ the entries $a_1, \ldots, a_l$, $l = m^2 s$, of all the matrices in a generating set for G. Since the field $\mathbb{Q}(a_1, \ldots, a_l)$ is finitely generated, the field K of algebraic numbers in it is finite over $\mathbb{Q}$. Since every eigenvalue λ of every matrix g in G is a root of a polynomial of degree at most m over K (namely, the characteristic polynomial), the minimal polynomial of g has bounded degree over $\mathbb{Q}$. This means that λ is a root of unity of bounded degree (because there are only finitely many cyclotomic polynomials of given degree: $\phi(m) \to \infty$ as $m \to \infty$, where $\phi(m)$ is Euler's function). Therefore, there is an integer n such that $g^n = 1$ for all g in G, and Burnside's theorem applies.

The question arises about what happens for non-linear groups. Is it true that every finitely generated group of finite exponent is finite? (This is the *Burnside problem.*) More generally, is it true that every finitely generated periodic group is finite?

The difficulty of these problems is emphasised by the fact that the answer is *yes* for soluble groups and many more general groups. The fact is that the commutator subgroup $[G, G]$ of a finitely generated periodic group is of finite index ($G/[G, G]$ is periodic abelian). Schreier's formula from 1.1.2 shows that $[G, G]$ is finitely generated in this situation. Induction on the solubility length now gives that the commutator subgroup is finite, and therefore so is G since $|G| = |G: [G, G]| \cdot |[G, G]|$.

The simplicity of the statement and the difficulty in solving the Burnside problem have induced the authors of Chandler and Magnus (1982) to compare its role in group theory with that of Fermat's Last Theorem in number theory. Undoubtedly, this circle of problems has a powerful influence on the theory of associative algebras, Lie algebras and other disciplines in contemporary algebra.

5.2. Residually Finite Periodic Groups. It was in 1964 that Golod discovered the negative solution of the "unrestricted" form of this problem.

Theorem. *For every $d \geqslant 2$ and every prime p there is an infinite group G on d generators such that every $(d-1)$-generator subgroup is a finite p-group.*

For $d = 2$, this result gives a finitely generated infinite periodic group. All of Golod's groups are residually finite. The construction rests on an important theorem about ideals in the free associative algebra $A_d = k[x_1, \ldots, x_d]$ of polynomials in noncommuting variables $x_1, \ldots, x_d$. This very general assertion states the following. For every k we choose m_k homogeneous polynomials of degree $k \geqslant 2$ in A_d and consider the ideal I generated by them; then, the factor-algebra A_d/I is infinite-dimensional if and only if m_k grows "comparatively slowly". The ideal I is chosen at a later stage so as to contain a power of every element a of A_d with zero free term: $a^{n(a)} \in I$. (It is easy to see that the elements a of A_d can be chosen so that $m_k \leqslant 1$ for all k, at the expense of a somewhat faster growth of the numbers $n(a)$.) Since a is nilpotent in A_d/I, the element $1 + a$ is invertible (with inverse $(1+a)^{-1} = 1 - a + a^2 - \cdots + (-a)^{n(a)-1}$). Moreover, the elements of this sort are of finite multiplicative order if k is the field of residues modulo p – this uses the binomial theorem. The fact that the group generated by the cosets $1 + x_i$ in A_d/I is infinite follows since A_d/I is infinite-dimensional; residual finiteness is a consequence of the homogeneity of I.

The deduction of the theorem for $d > 2$ is somewhat more complicated. It is clear from our short explanation that varying the choice of the generators of the ideal I can give rise to very diverse examples of residually finite p-groups.

The simplest example of a finitely generated infinite periodic group was published by Grigorchuk in 1980. In this example, the group G is generated by permutations a, b, c, d of the interval $[0, 1]$ with the points $k/2^n$ $(k, n = 0, 1, \ldots)$ removed.

The permutation a simply interchanges the two halves of the interval, preserving their direction, so that $a^2 = e$, the identity permutation. The permutations b, c, d are given in the same sort of way and are connected with the partition of the interval by the points $1/2, 3/4, 7/8, \ldots$. On $[0, 1/2]$, b acts like a does on $[0, 1]$, that is, b interchanges the intervals $[0, 1/4]$ and $[1/4, 1/2]$; on $[1/2, 3/4]$, b also acts like a, interchanging the two halves of this interval; on $[3/4, 7/8]$, b acts like the identity; thereafter, everything is repeated periodically (it acts like a on $[7/8, 15/16]$ *etc*). Similarly, c is defined periodically with sequence $a, e, a, a, e, a, \ldots$, and d periodically with sequence $e, a, a, e, a, a, \ldots$. The following relations follow from the definitions: $b^2 = c^2 = d^2 = e$, $bc = cb = d$, $bd = db = c$, $cd = dc = b$.

Obviously, the subgroup H consisting of the permutations that leave the two halves of the interval fixed contains the elements in which a occurs an even number of times. It is generated by the elements b, c, d, aba, aca, ada and can be embedded in the direct product $G \times G$ if one observes that b acts like a on the left half and like c on the right half; that is, $b = (a_l, c_r)$, and further $c = (a_l, d_r)$, $d = (e_l, b_r)$, $aba = (c_l, a_r)$, $aca = (d_l, a_r)$, $ada = (b_l, e_r)$. The group G is infinite since the projections of the proper subgroup H on the left-hand direct factor gives $\langle a_l, b_l, c_l, d_l \rangle$, which is isomorphic to the whole group G.

We show now that G is a 2-group, by induction on the length of a shortest expression of an element g of G in terms of the given generators. For this reason, nowhere in the expression for g are there successive occurrences of b, c, d; and the letter a stands at one end or the other (if $g = bg'c$, $b^{-1}gb$ has the same order as g and is equal to $g'cb = g'd$, which is a shorter word than g).

Thus, what we have to do is to show that $g = a * a * \cdots$ (the stars denote b, c or d) has 2-power order. If g is in H, then on replacing b by (a, c_r), c by (a, d_r), *etc*, we shorten the length of the components of $g = (g_l, g_r)$ in comparison with the length of the expression for g (indeed, every element of H has an even number of occurrences of a and so is a product of quadruplets $a * a * = (a * a) *$, and this means that both components of g have lengths which are shorter by a factor of two. The induction hypothesis now applies. If $g \notin H$, we go over to g^2, which is in H (doubling the length), and then represent g^2 as above (the lengths of the components are halved). In this situation, if d occurs in the expression for g, then d and ada must occur in the natural representation of g^2 in terms of the generators for H (this is easy to check using the fact that a occurs an odd number of times in the expression for g). But $d = (e_l, b_r)$, $ada = (b_l, e_r)$, so that both components of g^2 are strictly shorter than g. If d does not occur in g, but c does occur, the expressions for the two components of g^2 contains an occurrence of d, and this reduces the situation to the preceding one. Finally, if neither d nor c occurs, then b does and it can be used to produce an occurrence of c in both components of g^2. This completes the induction.

As is evident from the definition of the generators a, b, c, d, G preserves not only the decomposition of the interval $[0, 1]$ into halves, but also that into quarters, eighths *etc*. It follows that G is residually finite, the images being permutation groups of degrees 2, 4, 8, ... In fact, since G is a 2-group, G is residually a finite 2-group.

It would be hard to present a simpler example of a periodic non-locally finite group. Using this group and its descendants, Grigorchuk solved the Milnor problem about groups with intermediate growth (see 1.6.3), as well as some problems in neighbouring parts of algebra. The further development of mathematics, and not just of algebra, is thereby opened up; it could be that non-combinatorial counterexamples to the Burnside problem for groups of finite exponent (see 5.3 and 5.4 in Chapter 1) will be found using these methods. As a matter of fact, Grigorchuk found his example during his search for a non-amenable group.

Grigorchuk's group is close in substance to the periodic non-locally finite group of automata transformations constructed earlier (in 1972) by Aleshin (see Kargapolov and Merzlyakov (1982) for more details). Another and similar way of realizing periodic residually finite groups has been suggested by Sushchanskij, namely representations *via* isometries of the metric space of the p-adic integers, with its usual metric. The example of Golod is of a different nature from those just mentioned – it is given by a presentation and not by a representation.

Interest in residually finite periodic groups is stimulated, in particular, by the restricted Burnside problem (see 4.4). The negative solution for exponent n is equivalent to the existence of a finitely generated infinite residually finite group of exponent n. The fundamental result of Kostrikin described in 4.4 shows that no such example exists for prime n. If there is an example of prime-power exponent p^k, it could perhaps be located near the groups considered above. Unfortunately, those groups have elements of unbounded orders.

5.3. Groups of Finite Exponent. By now, when it has been solved in the negative in general (see 1.5.3), the Burnside problem has reverted to a sequence of problems, depending on the value of n. (And, generally speaking, on the number m of generators. However, up to the present, no situation has occurred where the problem has a positive solution for one pair (n, m_1) and a negative solution for another pair (n, m_2).)

As in 5.3 of Chapter 1, we denote by $\mathbb{B}(m, n)$ the largest group of exponent n on m generators, that is, the m-generator free group in the Burnside variety $\mathfrak{B}_n$ defined by the identity $x^n = 1$. Since groups of exponent 2 are commutative, $|\mathbb{B}(m, 2)| = 2^m$. Groups of exponent 3 are 2-Engel, as is easily checked using relations of the form $a^{\pm 2} = a^{\mp 1}$ and $y^{-1}x^{-1}y^{-1} = xyx$:

$$\begin{aligned}[][[x, y], y] &= xyx^{-1}yxy^{-1}x^{-1}y^{-1} = xyx^{-1}yx^2yx = xy(x^{-1}y)^2x \\ &= xy \cdot y^{-1}xx = x^3 = 1.\end{aligned}$$

This means that they are nilpotent (see 4.4) and locally finite. Levi and van der Waerden calculated the order of $\mathbb{B}(m, 3)$.

Theorem. $|B(m, 3)| = 3^{m+\binom{m}{2}+\binom{m}{3}}$.

Burnside himself proved the following result.

Theorem. *The order of every* 2*-generator group of exponent* 4 *is at most* 2^{12}.

For arbitrary generating number, the problem was solved by Sanov.

Theorem. *Groups of exponent* 4 *are locally finite.*

The fairly short and completey elementary proof M. Hall, Jr. (1959) of this theorem has exercised powerful influence, despite being short and elementary.

Since finite 2-groups are nilpotent (see 4.1), Sanov's theorem means that groups of exponent 4 are locally nilpotent. However, they need not be nilpotent, as the 3-Engel group of exponent 4 mentioned in 4.4 demonstrates. The solubility problem for groups of exponent 4, posed by Hall and Higman, was solved in the negative by Razmyslov using his method of α-functions on 2-words. Razmyslov used the same method to obtain his theorem on groups of exponent p (see 4.4), and also to solve an old problem in ring theory. The solution of the Hall-Higman problem is expounded in Bakhturin (1985).

Because the arguments in the proofs of these facts are pretty complicated, we shall give (briefly) a proof of the fact, also due to Razmyslov, that there is an insoluble locally finite group of exponent p^k for $k \geqslant 2$ and p odd.

Let A be an associative algebra with identity on the generators $x_1, x_2, \ldots$ over a field of characteristic $p > 0$ satisfying the relations $x_i w x_i = 0$, where w is any element of A, $i = 1, 2, \ldots$, and let G be the group generated by the elements $g_i = 1 + x_i$, $i = 1, 2, \ldots$. Since $g_i^{-1} = 1 - x_i$, G is generated *qua* semigroup by the elements $1 \pm x_i$. In order for G to have exponent p^k, A needs to be factored by an ideal contaning $g^{p^k} - 1 = (g - 1)^{p^k}$ for each g in G. An ideal I containing all elements of the form $(g - 1)^m$, $g \in G$, must also contain the element $f_{m,l} = ((1 + x_1)\ldots(1 + x_l) - 1)^m$, which (in view of the relation $x_i w x_i = 0$) is equal to $\sum_{y_1,\ldots,y_m} y_1 \ldots y_m$, where every y_i is a product of factors x_j with increasing indices, and the sum extends over all words $y_1, \ldots, y_m$ not contaning two identical letters. We denote the homogeneous components of leading degree in this polynomial by $f_m(x_1, \ldots, x)$. It is given by $f_m(x_1, \ldots, x_l) = \sum_{y_1,\ldots,y_m} y_1 \ldots y_m$, where $y_j = x_{i_1} \ldots x_{i_k}$ $(i_1 < \cdots < i_k)$, and the sum is over all $y_1, \ldots, y_m$ for which $y_1 \ldots y_m$ has exactly one occurrence of every letter x_i. The other homogeneous components are obtained from f_m on replacing x_i by x_{j_i}. If in addition to the defining relations imposed on A at the very beginning we add the relations $f_m(x_1, \ldots, x_{j_l}) = 0$ for all $j_1, \ldots, j_l$ and all l, the group generated by the elements $1 + x_i$ will have exponent $p^k \geqslant m$. Clearly, in A we have

$$g_i g_j g_i^{-1} g_j^{-1} = (1 + x_i)(1 + x_j)(1 - x_i)(1 - x_j) = 1 + [x_i, x_j] = 1 + x_i x_j - x_j x_i,$$

where $[\ ,\ ]$ denotes the ring commutator and $c(g_1, \ldots, g_k) = 1 + c(x_1, \ldots, x_k)$, where one and the same letter is being used to mean the group commutator and the analogous ring commutator. Thus, the group generated by the g_i, $i = 1, 2, \ldots$, is soluble if and only if the Lie algebra generated by $x_1, x_2, \ldots$ with respect to the operation $[a, b] = ab - ba$ is soluble.

The insolubility of this Lie algebra is established as follows. The algebra of 2×2 matrices over a field of characteristic $p > 0$ satisfies the identity $s_n = \sum_{(i_1,\ldots,i_n)} z_{i_1} \ldots z_{i_n}$ for $n = 4(p - 1) + 1$, where the sum is taken over all permutations on n symbols (this is not to be confused with the so-called "standard" identity, where the signs of the permutations come in). In fact, this identity need only be proved on the elements in a basis: since there are four elements in a basis, at least one arises p times, and therefore there is a factor $p!$ in front of the expression. We now take the free algebra B of the variety generated by the algebra of 2×2 matrices over an infinite field K, on generators $z_1, z_2, \ldots$; and we impose the relations $z_i u z_i = 0$ on it, just like for A. Since the identity $s_n = 0$ is satisfied, it follows that B also satisfies the relations $f_m(z_{i_1}, \ldots, z_{i_l}) = 0$ with $m = n$, which means that B is a homomorphic image of A. The insolubility of the Lie algebra generated by $z_1, z_2, \ldots$ under commutation is a consequence of the fact that the algebra of 2×2 matrices contains the simple algebra $sl_2(K)$.

The insolubility of $\mathbb{B}(\infty, 4)$ is proved using vastly subtler arguments. The resolution of the solubility problem was preceded by several articles; of these, we single out a result of Gupta and Newman, who showed that an m-generator group of exponent 4 is nilpotent of class at most $3m - 2$, and if $\mathbb{B}(m, 4)$ is nilpotent of class at most $3m - 3$ for any one value of m, then all groups of exponent 4 are soluble. Thus, we have:

Theorem. *The nilpotency class of* $\mathbb{B}(m, 4)$ *is precisely* $3m - 2$.

Although computers have been used to find the orders of $\mathbb{B}(3, 4)$ and $\mathbb{B}(4, 4)$, the structure of $\mathbb{B}(m, 4)$ for general m still awaits elucidation.

The structure of $\mathfrak{B}_6$ is simpler, even though the proof that it is locally finite is significantly more complicated that of Sanov's theorem. It is due to M. Hall, Jr.

Theorem. *Every group of exponent* 6 *is locally finite.*

Indeed, all such groups are soluble of length 3, since $\mathfrak{B}_6$ is minimal among the varieties containing the product varieties $\mathfrak{B}_3\mathfrak{B}_2$ and $\mathfrak{B}_2\mathfrak{B}_3$. Recently, Lysenok has found a short proof of the above theorem. It is based on ideas that are practically identical with those used to prove that groups of exponents 3 and 4 are locally finite.

That is all that is known in the positive direction on the restricted Burnside problem for finite exponent, which means that nothing new has happened since M. Hall wrote his book "Theory of Groups" in 1959. For the restricted problem (see 4.4), the answer is positive for all squarefree numbers and numbers of the form $4p$, where p is an odd prime. The solution comes from the theorems of Kostrikin and Sanov, together with the Hall-Higman theorem on the p-length of p-soluble groups and certain results of a classifying nature about finite simple groups.

By the Novikov-Adian theorem (see 1.5.3), $\mathbb{B}(m, n)$ is infinite for odd $n \geqslant 665$ and all $m \geqslant 2$, and all of its abelian subgroups are cyclic. Note for comparison that every infinite 2-group has an infinite abelian subgroup (see the next section). Of course, this proves nothing, but it is a clear indication that groups of exponent 2^k are nowhere nearly as tightly constructed as the groups $\mathbb{B}(m, n)$ for odd exponents. Currently, the Burnside problem for finite exponent is of most interest for the exponents 2^k, 5, $8 = 2^3$, $9 = 3^2$ and $12 = 3 \cdot 4$. We add at this point that Sanov showed that every group in which all elements are of order 1, 2, 3 or 4 is locally finite.

All known examples of infinite periodic groups – of finite exponent or not – are definined by infinite sets of defining relations. Thus, we return again to the problem which seems to have been originated by Novikov: do there exist finitely presented infinite periodic groups?

5.4. Locally Finite Groups. The local finiteness condition entered group theory long ago as one of the most natural finiteness conditions, even though it was not until the sixties that it was found to be stronger than periodicity. By now it can be said that free groups of finite exponent are more reminiscent of absolutely free groups than of finite or even of locally finite groups.

The class of locally finite groups is extremely extensive, and is closed under the operations of taken direct products, extensions, direct limits, subgroups and homomorphic images. The universal group of P. Hall is pivotal among the examples of locally finite groups.

Theorem. *There exists a countably infinite locally finite group G that embeds every countable locally finite group.*

The typical example of a simple locally finite group is $\mathrm{PSL}_n(k)$, $n \geqslant 2$, that is, the factor-group of the group of matrices of determinant 1 over an infinite field k by the (finite) subgroup consisting of the scalar matrices λE with $\lambda^n = 1$. If k can be written as the union of an ascending sequence of finite subfields k_i, then $\mathrm{PSL}_n(k) = \bigcup_{i=1}^{\infty} \mathrm{PSL}_n(k_i)$ is the union of an ascending chain of finite subgroups. Among other problems is the question as to whether every countable locally finite simple group G is the union of an ascending chain of finite simple groups[5]. It is easy to prove that G always has infinitely many nonisomorphic simple sections.

The minimum condition for subgroups (see 3.2) is a strong additional restriction under which the description of the structure of locally finite groups emerges. Clearly, min is inherent only in periodic groups, since the infinite cyclic group does not have the property. Thus, for (locally) soluble groups, min implies local finiteness (see the end of 5.1).

Every abelian group with min is the direct product of a finite group and finitely many quasicyclic groups. For, by the theorem in 1.2, we can assume that G is a reduced p-group, which means that $G^p \neq G$ if $G \neq \{1\}$, $G^{p^2} \neq G^p$ if $G^p \neq \{1\}$, *etc.* By min, $G^{p^n} = \{1\}$ for some n, and by Prüfer's theorem (see 1.3), G is the direct product of cyclic groups, and again by min, it is finite. In this same way, the divisible part is a direct product of quasicyclic groups and can have only finitely many factors. Conversely, every quasicyclic group satisfies the minimum condition, because all its proper subgroups are finite. The property min carries over to direct products (and indeed to extensions: if A and G/A satisfy min, so does G).

Definition. A Chernikov group is a finite extension of an abelian group satisfying the minimum condition.

The following result of Chernikov is the foundation for many subsequent generalizations and refinements.

Theorem. *Every soluble group, indeed every RN-group, with the minimum condition is a Chernikov group.*

Let us prove this in the case of a soluble group G. By min, G has a subgroup H which is minimal among those of finite index. It is normal in G, otherwise there would be a smaller subgroup of finite index of the form $H \cap gHg^{-1}$. Further, we pick any abelian normal subgroup M in H and show that it must be central in H. Since G is periodic, it is enough to establish that $M_n \subseteq Z(H)$ for the normal subgroups $M_n = \{a \in M | a^n = 1\}$. The group M_n is finite (by Prüfer's theorem, as above), which means that the centralizers of its elements are of finite

[5] This was disproved by A.E. Zalesskij and V.M. Serezhkin in 1976.

index in H, and by the choice of H it must actually be H, that is, $M_n \subseteq Z(H)$. To complete the proof, it is enough to establish that H is abelian. Assume that $Z(H) \neq H$. Since $H/Z(H)$ is soluble, we can find a nontrivial normal abelian subgroup of it, which must be central in $M/Z(H)$ by what has been proved already. But then, we can adjoin an element m of $M \setminus Z(H)$ to $Z(H)$, to obtain an abelian subgroup $\langle m, Z(H) \rangle$ which is normal in H. Once more, this subgroup must lie in $Z(H)$, and we have reached a contradiction.

Chernikov has proved that the same theorem holds for locally soluble groups under a slightly weaker assumption, namely the minimum condition on abelian subgroups.

Among the locally finite groups with the minimum condition, p-groups have particularly nice structure. They have been comprehensively studied by Chernikov. For example, such groups have nontrivial centres (like finite groups: see 4.1). To see this, we take a minimal normal subgroup A of a group G of this sort. It is finite, and so one can use the method of 4.1 to find central elements in it. Now G need not be nilpotent if it is not finite: an example is the semidirect product AB, where A is a group of type 2^∞, B is a cyclic group $\langle b \rangle$ of order 2 acting on A (see 2.1) by the rule: $bab^{-1} = a^{-1}$. Clearly, the centre of G is of order 2, and the same is true for the factors of the upper central series of G.

There is a great gulf between these groups and certain examples of finitely generated groups with the minimum condition (see 5.4 of Chapter 1), which show that by no means all p-groups with min are Chernikov (this was a problem that Chernikov himself had posed). No "intermediate" groups with min are visible at the moment. Following the modern fashion, we could describe Chernikov groups as tame, and other groups with min as wild. The tame groups with max are the finite extensions of polycyclic groups (see 3.2). Compared with the Ol'shanskij examples (see 1.5.4), the – still hypothetical – residually finite non-tame groups with max could be considered as being intermediate. The same examples give the negative answer to Shmidt's problem (see 1.5.4) in its full generality; however, for locally finite groups the solution is positive and can be found in papers by Kargapolov, and Hall and Kulatilaka.

Theorem. *If all proper subgroups of a locally finite group G are finite, then G is cyclic or quasicyclic.*

Even more has been proved: every infinite locally finite group has an infinite abelian subgroup. Essential use is made of the Feit-Thompson theorem on the solubility of groups of odd order, and the Brauer-Fowler theorem stating that there are only finitely many nonisomorphic finite simple groups with given centraliser of an involution (element of order 2).

We shall show next that every finite subgroup of an infinite 2-group is strictly smaller than its normaliser. Thus, let H be a finite subgroup with $N(H) = H$. Since not every finite subgroup is contained in H, we can choose a finite subgroup K such that $D = K \cap H$ is maximal for those K with $K \nsubseteq H$. Since D is a proper subgroup both of K and of H, it is different from its normalisers in K and in H. We choose elements a and b of these normalisers respectively having order

2 modulo D, and consider the subgroup $\langle a, b, D\rangle$ generated by them together with D. This subgroup is finite, since it is generated modulo D by two involutions, and thus the factor-group modulo D is finite. We now have the following: $\langle a, b, D\rangle$ is a finite group not contained in H, whereas $\langle a, b, D\rangle \cap H$ is larger than D since it contains the element b outside D. This contradiction establishes our assertion. It follows that every infinite 2-group has an infinite locally finite subgroup: it is enough to take the union of a chain $H_1 \subseteq H_2 \ldots$ of finite subgroups such that $H_{i+1} \subseteq N(H_i)$ and $H_{i+1} \neq H_i$. From the theorem mentioned just now (actually, this is an enormously simpler fact!) it follows that every infinite 2-group has an infinite abelian subgroup.

A decisive part in the foregoing argument is the fact that two elements of order 2 in a periodic group generate a finite subgroup, an assertion that has no analogue for p-groups with $p \neq 2$. It is typical for theorems of this general type. We ask now: how can we prove Shmidt's theorem stating that every 2-group with min is locally finite? One looks at a minimal counterexample, and let G be the factor-group by a maximal normal subgroup. Then prove that every two different maximal subgroups M_1 and M_2 have trivial intersection. Choose involutions $a_1 \in M_1$, $a_2 \in M_2$. They generate a finite subgroup, which is contained in a maximal subgroup M. But $a_1 \in M \cap M_1$, $a_2 \in M \cap M_2$, so that $M = M_1 = M_2$. This means that G has a unique maximal subgroup M, which is therefore normal. In its turn, this contradicts the simplicity of G.

As even stronger influence on the achievements of finite group theory was exerted by Shunkov's solution of the problem of Chernikov in the class of all locally finite groups. (This result was obtained at the same time by Kegel and Wehrfritz: their paper was based on previous articles by Shunkov.)

Theorem. *Every locally finite group with the minimum condition is Chernikov.*

On examining a hypothetical minimal counterexample to this assertion, it is not difficult to show that G is a simple group, and even the union of finite simple groups. A decisive moment in the proof is the application of Shunkov's characterization of the simple group $\mathrm{PSL}_2(k)$ over an infinite locally finite field k of odd characteristic.

Further sharpenings and generalisations were obtained by Shunkov, at the expense of weakening the requirement of min (to abelian subgroups only) and extending the class of locally finite groups to periodic groups with some additional requirements. The result just mentioned is one of the central planks of the theory of locally finite groups.

Another intensively-studied class of groups is that consisting of groups with finite conjugacy classes of elements (the FC-*groups*). If G is a periodic group of this type, then every finite subset lies in a finite normal subgroup (G is *locally normal*). To see this, we shall show that a conjugation-closed finite subset $g_1, \ldots, g_m$ of a periodic group generates a finite (and obviously normal) subgroup. If N is the least common multiple of the orders of $g_1, \ldots, g_m$, then every element of $\langle g_1, \ldots, g_m\rangle$ can be written in the form $g_{i_1} \ldots g_{i_k}$, where $k \leqslant mN$. For, if $k > mN$, at least one element, g_j say, occurs more than N times. Shifting all the g_j to the first position using the formula $g_i g_j = g_j(g_j^{-1} g_i g_j) = g_i g_s$ (valid since our generating set is conjugation-closed), we get that $g_{i_1} \dots g_{i_k} = g_j^N g_{s_1} \dots g_{s_t}$, where $t = k - N$. Since $g_j^N = 1$, the expression for our element diminishes. The whole theory has been developed around locally normal groups; typical examples of these are the direct products of finite groups. Every locally normal group has the FC-property, of course, and the central factor-group $G/Z(G)$ of an arbitrary FC-group is locally normal (Baer). An example of a book devoted to the theory of FC-groups is Gorchakov (1978).

At first sight it would appear natural to consider the conjugacy problem for the maximal p-subgroups of locally finite groups. However, this problem has a positive answer in very few cases. A general theorem due to Asar and Hartley states the following. If for some p every countable subgroup of a locally finite group has only countably many maximal p-subgroups, the maximal p-subgroups are conjugate. Of course, the converse holds. Take the simplest kind of locally finite group: the direct product of countably infinitely many isomorphic copies of the symmetric group S_3. Conjugate subgroups must coincide in all but the finitely many places where the conjugating element has non-trivial coordinates. However, we can take the subgroup generated, shall we say, by the strings that have (12) in all places where they are non-trivial, and similarly for (13). However, in this group, and more generally in all locally normal groups, the maximal p-subgroups are locally conjugate: that is, each is carried to the other by an automorphism that acts like conjugation on every finite set of elements.

References*

Adian, S.I. (1975): The Burnside Problem and Identities in Groups. Nauka, Moscow, Zbl.306.20045. English transl.: Ergebn. Math. Grenzgeb., Z. Folge, 95. Springer-Verlag, Berlin Heidelberg New York 1979, Zbl.417.20001

Bakhturin, Yu.A. (1984): Identities in Lie Algebras. Nauka, Moscow, Zbl.571.17001. English transl.: VNU Science Press, Utrecht 1987

Bakhturin, Yu.A., Ol'shanskij, A.Yu. (1988): Identities. Itogi Nauki Tekhn., Ser. Sovrem. Probl. Mat., Fundam. Napravleniya 18, 117–240, Zbl.673.16001. English transl. in: Encycl. Math. Sci. 18. Springer-Verlag, Berlin Heidelberg New York 1991

Baumslag, G. (1971): Lectures in Nilpotent Groups. C.B.M.S. Reg. Conf. Ser. 2. Am. Math. Soc., Providence R.I., Zbl.241.20001

Brown, K.S. (1982): Cohomology of Groups. Springer-Verlag, New York Berlin Heidelberg, Zbl.584.20036

* For the convenience of the reader, references to reviews in Zentralblatt für Mathematik (Zbl.), compiled using the MATH database, and Jahrbuch über Fortschritte der Mathematik (Jbuch) have, as far as possible, been included in this bibliography.

Burnside, W. (1911): The Theory of Groups of Finite Order. Cambridge University Press, Cambridge. Reprint: Dover, New York 1955, Zbl.64.251

Chandler, B., Magnus, W. (1982): A History of Combinatorial Group Theory: A Case Study in the History of Ideas. Springer-Verlag, Berlin Heidelberg New York, Zbl.498.20001

Chernikov, S.N. (1980): Groups Having Subgroup Systems with Given Properties. Nauka, Moscow (Russian), Zbl.486.20020

Cohen, D.E. (1972): Groups of Cohomological Dimension One. Springer-Verlag, New York Berlin Heidelberg, Zbl.231.20018

Crowell, R.H., Fox, R.H. (1963): Introduction to Knot Theory. Ginn and Co., Boston New York Chicago etc., Zbl.126.391

Dicks, W. (1980): Groups, Trees and Projective Modules. Springer-Verlag, Berlin Heidelberg New York, Zbl.427.20016

Fuchs, L. (1973): Infinite Abelian Groups. Vols 1 and 2. Academic Press, New York London, Zbl.209.55 and Zbl.257.20035

Gorchakov, Yu.M. (1978): Groups with Finite Classes of Conjugate Elements. Nauka, Moscow (Russian), Zbl.496.20025

Grossman, I., Magnus, W. (1964): Groups and their Graphs. Random House, L.W. Singer Co., New York, Zbl.123.248

Gruenberg, K.W. (1970): Cohomological Topics in Group Theory. Springer-Verlag, Berlin Heidelberg New York, Zbl.205.327

Gupta, N.D. (1987): Free Group Rings. Am. Math. Soc., Providence R.I., Zbl.641.20022

Hall, M. Jr. (1959): The Theory of Groups. Macmillan, New York, Zbl.84.22

Hall, P. (1957): Nilpotent Groups. Queen Mary College Mathematical Notes, London, 1969, Zbl.211.342

Jacobson, N. (1962): Lie Algebras. Interscience Publ., New York London, Zbl.121,275

Kargapolov, M.I., Merzlyakov, Yu.I. (1982): Elements of the Theory of Groups. Nauka, Moscow, Zbl. 256.2002. English transl.: Springer-Verlag, New York Berlin Heidelberg 1979

Kegel, O.H., Wehrfritz, B.A.F. (1973): Locally Finite Groups. North-Holland Co., Amsterdam London, Zbl.259.20001

Kostrikin, A.I. (1986): Around Burnside. Nauka, Moscow, Zbl.624.17001. English transl.: Ergebn. Math. Grenzgeb., 3. Folge, 20. Springer-Verlag, Berlin Heidelberg New York 1989, Zbl.702.17001

Kurosh, A.G. (1967): Theory of Groups. Nauka, Moscow 1967, Zbl.189.308. German transl.: Akademie-Verlag, Berlin 1972, Zbl.235.20001

Lyndon, R.C., Schupp, P.E. (1977): Combinatorial Group Theory. Springer-Verlag, Berlin Heidelberg New York, Zbl.368.20023

MacLane, S. (1963): Homology. Springer-Verlag, Berlin Heidelberg New York, Zbl.133.265

Magnus, W., Karras, A., Solitar. D. (1966): Combinatorial Group Theory. Interscience Publ., New York London Sydney, Zbl.138.256

Mal'tsev, A.I. (1970): Algebraic Systems. Nauka, Moscow, Zbl.223.08001. English transl.: Grundl. Math. Wiss. 192. Springer-Verlag, New York Berlin Heidelberg 1973, Zbl.266.08001

Massey, W.S. (1967): Algebraic Topology: An Introduction. Harcourt and Brace, New York, Zbl.153.249

Neumann, H. (1967): Varieties of Groups. Ergebn. Math. Grenzgeb., 2. Folge, 37. Springer-Verlag, Berlin Heidelberg New York, Zbl.251.20001

Ol'shanskij, A.Yu. (1989): The Geometry of Defining Relations in Groups. Nauka, Moscow, Zbl.676.20014. English transl.: Kluwer Publ., Dordrecht 1991

Plotkin, B.I. (1966): Groups of Automorphisms of Algebraic Systems. Nauka, Moscow. Zbl.192,115. English transl.: Wolters-Noordhof, Groningen 1972

Plotkin, B.I., Vovsi, S.M. (1983): Varieties of Group Representations. Zinatne, Riga (Russian), Zbl.527.20004

Pontryagin, L.S. (1954): Topological Groups. Gostekhizdat, Moscow, Zbl.58.260. English transl.: Gordon and Breach, New York London Paris 1966

Razmyslov, Yu.P. (1989): Identities of Algebras and their Representations. Nauka, Moscow (Russian), Zbl.673.17001

Robinson, D.J.S. (1982): A Course in the Theory of Groups. Springer-Verlag, New York Berlin Heidelberg, Zbl.483.20001
Segal, D. (1983): Polycyclic Groups. Cambridge University Press, Cambridge, Zbl.516.20001
Shmidt, O.Yu. (1916): Abstract Group Theory. Kiev, 1916; Collected Works. Matematika, Moscow 1959, pp. 17–175 (Russian), Zbl.87.255
Stallings, J. (1971): Group Theory and Three-Dimensional Manifolds. Yale University Press, New Haven London, Zbl.241.57001

II. Linear Groups

A.E. Zalesskij

Translated from the Russian
by J. Wiegold

Contents

Introduction

It is hardly necessary to expatiate here on the widespread and manifold applications of group theory. For all that, it must be emphasised that applications are always associated with realisations of groups as groups of transformations (essentially, that is, as groups of symmetries) of some mathematical system or other. Without doubt, the most important types of transformation groups are the groups of linear transformations, that is, the *linear groups*. Their significance in the natural sciences was appreciated at the very dawn of the development of group theory. One of the earliest and most impressive instances of this is the classification of crystallographic groups (1890).

The aim of this survey is to give an introduction to the theory of linear groups that is accessible to non-specialists, and within that aim to cover in brief the basic ideas and the contemporary content of the subject. We shall be concerned with the achievements of the general theory of linear groups, which develops methods for investigating whole classes of groups. Problems about structure and the actions of individual groups will remain outside our line of sight, by and large. This does not mean that the author fails to appreciate the importance of knowing about concrete groups. Nevertheless, the systematisation of material of that sort is not in tune with the aims of this survey.

Chapter 1 introduces the principal objects in the theory of linear groups. Chapter 2 is devoted to a presentation of the apparatus of the theory, and Chapter 3 covers the content of the modern theory. In reality, the three chapters pursue the same aim: they illuminate the main theme from different points of view. We emphasise here that our survey cannot pretend to be exhaustive, because it is absolutely inconceivable that we could touch on the whole of the vast collection of concrete results that have been obtained up to the present. Rather, it should be viewed as a sort of guide-book to the massive modern theory.*

The level of detail in our exposition is normally such as to orient the non-expert reader who has encountered problems connected with linear groups.

Linear group theory is very closely connected with the theories of Lie groups, algebraic groups and group representations; with algebraic K-theory and the theories of bilinear and quadratic forms. Although we had not intended to expound these themes, representing as they do the most fundamental thrusts in algebra, we have been forced to pay significant attention to them.

We shall dwell a little on certain other, albeit minor, facets. The well-known correspondence between matrices and linear maps enables us to expound the

---

* The text of this survey was written in 1986. A lot of new results have been obtained during the last years. A number of minor additions have been made to reflect the most important among them, but much less has been added than I would have liked. In particular, quite new technique of applying profinite and pro-p-groups theory to linear groups is not explained here. The reader should consult the recent book of of Dixon, du Sautoy, Mann and Segal (1991).

theory in the language of matrices as well as in that of linear maps. The reader not seduced by modern algebra as it appears to us will find the language of matrix theory easier to accept. Thus, we have given it preference in the first chapter, where all readers can acquaint themselves with the basic objects of the theory. In the sequel, when we are dealing with the main methods and results in the theory, preference is given to the language of linear maps.

The algebraical reader will find it totally natural that many constructions and results are dealt with for arbitrary ground fields. Those who are unfamiliar with this degree of generality can replace the arbitrary field by one with which they are familiar. For example, the rational field will do for a field of characteristic zero in general, and the field of complex numbers for an algebraically closed field of characteristic zero. The loss of generality arising thereby is not all that important. Fields of positive (prime, finite) characteristic are less familar to the non-algebraist; however, they too function very successfully, for example in coding theory. The role of fields of prime characteristic in modern algebra, and in the theory of linear groups in particular, is so great that it would be quite wrong to ignore results involving them.

In the main, the references comprise survey books and articles that might be of use to the reader desirous of closer familiarity with whatever themes interest him. Moreover, many results will carry reference to author's surname and year of publication; these data are usually enough to locate the exact coordinates of an article in the bibliographies of the survey articles in our reference list or in the indexes of reviewing journals.

The requirements in the way of preparation on the part of the reader are minimal, especially in §§ 1–3 of the first chapter and §§ 1, 2 of the second. All the same, some readers might feel that they need to orient themselves more fully in the elements of the general theory of algebraic systems such as groups, fields, rings, modules *etc.* In that case, we recommend the introductory course to be found in Shafarevich (1985), or some other, similar, source.

It is convenient to agree on some notation at this point. The symbols $\mathbb{Q}$, $\mathbb{R}$, $\mathbb{C}$ denote the fields of rational, real and complex numbers respectively; $\mathbb{Z}$ is the ring (or set) of integers. $\mathbb{N}$ the set of natural numbers. The number of elements in a set M is denoted by $|M|$. For a matrix A, its trace, determinant and transpose are denoted by $\operatorname{tr}(A)$, $\det A$ and ${}^tA$ respectively; E_n is the $n \times n$ identity matrix of degree n (the suffix n may be omitted if it is clear from context). The set of all $n \times n$ matrices with coefficients in a field (or ring) R is denoted by $M_n(R)$, and the group of all invertible matrices by $\mathrm{GL}_n(R)$. If P is a field, its characteristic is written char P.

Internal references are given in three versions; for example, 1.2.5 denotes Chapter 1, § 2, subsection 5; 3.7 means § 3, subsection 7 in the current chapter.

Chapter 1
The Main Types of Linear Groups

§ 1. Simplest Examples of Matrix Groups

The examples given below are among the most basic. In the majority of them, it is not necessary to require that the matrix entries lie in a field (though that is assumed throughout the bulk of the work). In order to achieve the natural level of generality, it is better for the present to consider matrices over an arbitrary ring with identity. In this section, K denotes a ring with identity element 1.

1.1. The General Linear Group. Let n be a natural number. The *general linear group* $\mathrm{GL}_n(K)$ of degree n over the ring K consists of all $n \times n$ invertible matrices with entries in K. All other matrix groups of degree n over K are subgroups of it. Of course, $\mathrm{GL}_1(K)$ is just the group K^0 of invertible elements (*units*) of K. If K is a field and V a finite-dimensional vector space over K, the group $\mathrm{GL}(V)$ of all invertible linear maps is also called the general linear group. If $\dim V = n$, $\mathrm{GL}(V)$ is isomorphic with $\mathrm{GL}_n(K)$.

1.2. The Group of Scalar Matrices. It consists of the matrices $\mathrm{diag}(k, \ldots, k)$, with $k \in K^0$; we denote it by $Sc_n(K)$.

1.3. The Group of Diagonal Matrices. It comprises the matrices of the form $\mathrm{diag}(k_1, \ldots, k_n)$, where $k_1, \ldots, k_n \in K^0$; it is denoted by $D_n(K)$.

1.4. Upper and Lower Triangular Groups. The first, denoted by $T^n(K)$, consists of all invertible matrices with zeros below the main diagonal. The second is defined symmetrically and is denoted by $T_n(K)$.

1.5. Upper and Lower Unitriangular Groups. These are defined as the groups of upper (lower) triangular matrices with the identity along the main diagonal; the notation is $Y^n(K)$ and $Y_n(K)$.

We note that matrices like this are always invertible. Associating each upper-triangular matrix with its diagonal defines a homomorphism of $T^n(K)$ onto $D_n(K)$ with kernel $Y^n(K)$. The analogous remark applies to $T_n(K)$ and $Y_n(K)$.

1.6. Groups of Monomial Matrices. A matrix is said to be *monomial* if each row and each column contains exactly one non-zero entry. The *full monomial group* $\mathrm{Mon}(K)$ consists of all the invertible monomial matrices. Groups contained in $\mathrm{Mon}(K)$ are called *groups of monomial matrices* or simply *monomial groups*. The monomial matrices whose entries are zero or the identity form the group Π_n of *permutation matrices*. Clearly, Π_n is isomorphic with the *symmetric group* Sym_n, that is, the group of all permutations of n symbols. If for K we take the ring $M_m(R)$ of $m \times m$ matrices over R, Mon(K) can be represented by matrices of degree mn over R; these groups are called *groups of block-monomial matrices.*

The idea of constructing matrices of high degree from matrices of lower degree enables us to introduce certain intuitive examples of matrix groups.

1.7. Groups of Block Matrices. Every ordered r-tuple $[\lambda] = \{n_1, \ldots, n_r\}$ of natural numbers can be associated with the *group* $D_{[\lambda]}(K)$ *of block-diagonal matrices* of degree $n = n_1 + \cdots + n_r$; it consists of all the invertible matrices with zeros outside the corresponding blocks. As an abstract group, it is isomorphic with the direct product of the groups $\mathrm{GL}_{n_1}(K), \ldots, \mathrm{GL}_{n_r}(K)$. The *group of upper block-triangular matrices*, $T^{[\lambda]}(K)$, consists of the invertible matrices with zeros in positions below the diagonal blocks of dimensions $n_1, \ldots, n_r$ (disposed from above to below according to the orders of their suffixes). Associating each block-triangular matrix with its "block" diagonal defines a homomorphism $T^{[\lambda]}(K) \to D_{[\lambda]}(K)$. The kernel of this homomorphism is the group of *upper block-unitriangular matrices.* In other words, this is a subgroup of the block-triangular group formed from matrices with identity matrices in the diagonal blocks.

1.8. Net Groups. We shall divide matrices into cells by a fixed set of vertical and horizontal lines symmetric with respect to the diagonal. In multiplying such matrices, we note that, in a certain sense, multiplication is consonant with the decomposition of the matrices into cells. This observation is a reflection of the following more sophisticated example.

A set of additive subgroups σ_{qr} of K $(q, r = 1, \ldots, n)$ is said to be a net of degree n if $\sigma_{qr}\sigma_{rs} \subset \sigma_{qs}$ for all q, r, s. We denote by M_σ the set of all matrices over K whose (q, r)-entires lie in σ_{qr}. Clearly, M_σ is a subring of $M_n(K)$. The *net group* G_σ is the largest subgroup contained in the set of all matrices of the form $E_n + A$ with $A \in M_\sigma$. The groups in 1.7 are examples of net groups.

1.9. The Special Linear Group. This is defined only for commutative rings K, and consists of all matrices with determinant 1. Notation: $\mathrm{SL}_n(K)$.

§ 2. Some Linear Group Constructions

The considerations of § 1 make one think about simple devices for producing new examples of linear groups out of known ones. We keep to the notation of § 1.

2.1. Similarity of Matrix Groups. Let H be a subgroup of $\mathrm{GL}_n(K)$. Every group of the form $H_1 = AHA^{-1}$ (for A in $\mathrm{GL}_n(K)$) is said to be *similar* to H; H_1 is also said to be *conjugate* to H by A.

2.2. Components. A group $H \subset \mathrm{GL}_n(K)$ is said to be *reducible* if it is similar to a subgroup L of $T^{[\lambda]}(K)$ for some $[\lambda] = \{n_1, \ldots, n_r\}$ in which $n_1 + \cdots + n_r = n$, $r > 1$. Otherwise, H is said to be *irreducible*.

Suppose that $AHA^{-1} = L \subset T^{[\lambda]}(K)$. With each matrix h in H we associate the submatrix of L positioned in the j-th diagonal block of $T^{[\lambda]}(K)$, and get a group homomorphism. The image H_i of this homomorphism in $\mathrm{GL}_{n_i}(K)$ is called a *component* of H. If H_i is irreducible, we speak of an *irreducible component*.

It is easy to see that for every subgroup H of $\mathrm{GL}_n(K)$ there is a decomposition $n = n_1 + \cdots + n_r$ of n such that all the components $H_i \subset \mathrm{GL}_{n_i}(K)$ are irreducible. Therefore, every group has irreducible components.

If H is a group similar to a subgroup L of $D_{[\lambda]}(K)$, with $r > 1$ as above, then H is said to be *decomposable* (and indecomposable otherwise). If $[\lambda]$ can be chosen so that the components H_i are irreducible, H is said to be a *completely reducible* linear group.

2.3. Generation of Linear Groups. Let S be any subset of $\mathrm{GL}_n(K)$. We consider the set of all products of matrices from S and their inverses. This gives a group, and we shall denote it by $\langle S \rangle$. If $H = \langle S \rangle$, S is called a generating set for H.

Although different sets S may generate identical groups $\langle S \rangle$, the thought arises immediately that the collection of linear groups is enormous, and that the problem of describing them completely is scarcely realistic. As we shall see, this suspicion is justified. It means that the theory must be developed in such a way as to form a guide through the chaos of over-numerous examples.

2.4. Direct Sums. Let H_i be a subgroup of $\mathrm{GL}_{n_i}(K)$, for $i = 1, \ldots, r$. Set $[\lambda] = \{n_1, \ldots, n_r\}$ and $n = n_1 + \cdots + n_r$. We consider the subgroup H of $D_{[\lambda]}(K)$ defined as follows: a matrix A in $D_{[\lambda]}(K)$ lies in H if and only if its i-th block A_i lies in H_i. It is easy to see that H is abstractly isomorphic to the direct product of the H_i. Note that renumbering the H_i gives a similar group. We shall write $H = \oplus H_i$ and call H the *direct sum* of the H_i.

2.5. Triangular Sums. With the notation of 2.4, we consider the set T^H of all matrices of the form $A + B$, where A is in H and B is a matrix with zeros on and below the diagonal blocks of H. It is easy to see that T^H is the inverse image of H under the homomorphism $T^{[\lambda]}(K) \to D_{[\lambda]}(K)$ (see 1.7), so that T^H is a group. We shall call it the *triangular sum* of the H_i.

2.6. Matrix Wreath Products. Let H be a linear group, $H \subset \mathrm{GL}_n(P)$, and Γ a subgroup of the group Sym_r of permutations on r symbols. We realise Γ as a subgroup of the group Π_r from 1.6, and then as a subgroup Γ_1 of the subgroup Π_{nr} of $\mathrm{GL}_{nr}(K)$ obtained by "exaggerating" the matrix entries of Π_r by a factor of n. The *matrix wreath product* $H \wr \Gamma$ is the subgroup of $\mathrm{GL}_{nr}(K)$ generated by the groups $H_r = H \oplus \cdots \oplus H$ (r times) and the realization Γ_1 of Γ as a subgroup of Π_{nr}. We note that H_r is normal in $H \wr \Gamma$, so that $H \wr \Gamma = H_r\Gamma_1$ is a (semidirect) product.

From now on, until the end of the section, the underlying ring K will always be assumed to be commutative.

2.7. Kronecker Products. Let $A = (a_{ij})$ and $B = (b_{ij})$ be square matrices of degrees m and n respectively. The *Kronecker product* $A \otimes B$ is the matrix

$$\begin{bmatrix} a_{11}B & \dots & a_{1m}B \\ \dots & \dots & \dots \\ a_{m1}B & \dots & a_{mm}B \end{bmatrix},$$

where each occurrence $a_{ij}B$ means the usual multiplication of a matrix by a scalar from K. Thus, the entries of $A \otimes B$ are the products $a_{ij}b_{kl}$, disposed in the way that is clear from what was said above. It is not hard to check that $A \otimes B$ and $B \otimes A$ are similar.

Let A_1 and A_2 be $m \times m$ matrices and B_1 and B_2 $n \times n$ matrices. Properties of matrix multiplicaiton give at once that $(A_1 \otimes B_1)(A_2 \otimes B_2) = A_1A_2 \otimes B_1B_2$. It follows easily that the matrices $A \otimes B$, with $A \in \mathrm{GL}_m(K)$ and $B \in \mathrm{GL}_n(K)$, form a group. Moreover, if A runs over a subgroup G_1 of $\mathrm{GL}_m(K)$ and B over a subgroup G_2 of $\mathrm{GL}_n(K)$, the Kronecker products $A \otimes B$ form a subgroup of $\mathrm{GL}_{mn}(K)$: it is called the *Kronecker product* of the matrix groups G_1 and G_2 (sometimes the term "tensor product" is used).

There is nothing to prevent us from looking at Kronecker products with arbitrarily many factors. It might look as though the Kronecker product of linear groups is abstractly isomorphic to their direct product. However, this is not the case. In fact, the kernel of the natural homomorphism $\mathrm{GL}_{n_1}(K) \times \dots \times \mathrm{GL}_{n_r}(K) \to \mathrm{GL}_{n_1}(K) \otimes \dots \otimes \mathrm{GL}_{n_r}(K)$ from the direct product of the general linear groups $\mathrm{GL}_{n_i}(K)$ to their Kronecker product consists of the elements $(a_1E_{n_1}, \dots, a_rE_{n_r})$, $a_i \in K$, such that $a_1 \dots a_r = 1$. Clearly, the kernel of the homomorphism from the direct product of arbitrary groups $G_i \subset \mathrm{GL}_{n_i}(K)$ to their Kronecker product is the set of all matrices $(a_1E_{n_1}, \dots, a_rE_{n_r})$ such that $a_iE_{n_i} \in G_i$ and $a_1 \dots a_r = 1$.

2.8. Linear Representations of Groups. A number of important linear group constructions are connected with the representation theory of groups. We recall some terminology and facts from representation theory; see Curtis and Reiner (1962).

Let G be an abstract group. A homomorphism $G \to \mathrm{GL}_n(K)$ is called a *linear representation* of G over the ring K; the number n is called the dimension, or the degree, of the representation. In the same way one speaks of representations of G in $\mathrm{GL}(V)$, where V is some vector space over a field P, which could be infinite-dimensional. However, when we mention group representations, we shall always have in mind finite-dimensional representations. On the rare occasions when we encounter infinite-dimensional representations, we shall stipulate this explicitly.

The image $\phi(G)$ of a linear representation ϕ of a group G is a linear group. Thus, the study of an individual representation is virtually identical with the study of a linear group. This fact is reflected in the terminology: a representation ϕ is said to be *reducible*, *irreducible*, *decomposable*, *indecomposable*, *monomial*, *imprimitive*, *primitive etc* if the same is true for the linear group $\phi(G)$.

The homomorphism of G to the identity group in $\mathrm{GL}_n(P)$ is called the *trivial representation*. Two representations $\phi_i\colon G \to \mathrm{GL}_{n_i}(P)$, $i = 1, 2$, are said to be *equivalent* if $n_1 = n_2$ and $\phi_1(g) = A\phi_2(g)A^{-1}$ for some invertible matrix A and all g in G. Otherwise, the representations are said to be *inequivalent*. The groups $\phi_i(G)$ can be identical even when the ϕ_i are inequivalent: the simplest example is where G is a cyclic group of order more than 2, ϕ_1 is a faithful representation in $\mathrm{GL}_1(\mathbb{C})$ and $\phi_2(g) = \phi_1(g^{-1})$ for all g in G.

Representations ϕ_1 and ϕ_2 are said to be *contragredient or dual* if ϕ_2 is equivalent to the representation $g \to {}^t\phi_1(g^{-1})$, $g \in G$.

In connection with linear groups, we shall sometimes speak of identification representations (that is, we separate the linear groups from its matrices and then re-identify them; this strange operation is sometimes forced upon us).

The *character* of a representation is the function $g \to \operatorname{tr}\phi(g)$ (the trace of the representing matrix). Characters of equivalent representations are identical. Remarkably, there is a converse: if the ground field P is algebraically closed, then two irreducible representations with identical characters are equivalent.

2.9. Direct Sums of Representations. Let $\phi_i\colon G \to \mathrm{GL}_{n_i}(K)$ $(i = 1, \ldots, r)$ be representations of a group G. The direct sum $\phi = \oplus\, \phi_i$ of these representations associates elements g of G with the block-diagonal matrices $\operatorname{diag}(\phi_1(g), \ldots, \phi_r(g))$. We emphasise that the group $\phi(G)$ is not the direct sum $\oplus\phi_i(G)$ of the linear groups $\phi_i(G)$ (see 2.4).

2.10. Tensor Products of Representations. Let ϕ_i be as in 2.9. The tensor product $\psi = \otimes\phi_i$ associates the element g of G with its image under the natural composition of homomorphisms:

$$g \xrightarrow{(\varphi_1, \ldots, \varphi_r)} \mathrm{GL}_{n_1}(K) \times \cdots \times \mathrm{GL}_{n_r}(K) \to \mathrm{GL}_{n_1}(K) \otimes \cdots \otimes \mathrm{GL}_{n_r}(K)$$
$$\to \mathrm{GL}_{n_1 \ldots n_r}(K)$$

(see 2.6). As in 2.9, $\psi(G)$ will usually be different from the Kronecker product of the groups $\phi_i(G)$ (taken independently), but the matrix $\psi(g)$ is the Kronecker product $\phi_1(g) \otimes \cdots \otimes \phi_r(g)$. If all the ϕ_i are equivalent, ψ is sometimes referred to as a tensor power of ϕ_1.

2.11. Adjoint Representations. At this point, it is convenient to assume that K is a field, P say. Let $\phi\colon G \to \mathrm{GL}_n(P)$ be a representation of a group G. With each element g of G we associate the linear map $\pi(g)$ of the vector space $M_n(P)$ given by the formula $\pi(g)A = \phi(g)A\phi(g^{-1})$, $A \in M_n(P)$. It is clear that π is a linear representation of G; we call it the representation *adjoint* to ϕ, and denote it by $\operatorname{Ad}\phi$.

(Note that the definition of adjoint representation used in the theories of Lie groups and algebraic groups is somewhat different from this one.)

One can also consider the representation $A \to \phi(g)A$ ($A \to A\phi(g^{-1})$ respectively) by left (right, respectively) multiplications. However the representation by left multiplications is the direct sum of n copies of ϕ (realised in the matrix columns) and so gives nothing essentially new. The other case produces a direct sum of representations contragredient with ϕ.

The reader who has mastered the elements of module theory can easily extend the construction of Ad ϕ to the case where K is a commutative ring.

2.12. Induced Representations. Let H be a subgroup of finite index r in some group G, and $\phi: H \to \mathrm{GL}_n(K)$ a linear representation. Let $g_1 = 1, g_2, \ldots, g_r$ be a left transversal of G modulo H. We consider matrices of degree nr, which are to be thought of as written out in square blocks (submatrices) of degree n. With each element g of G we associate the matrix $\psi(g)$ of degree nr in which the (i, j)-block ($1 \leqslant i, j \leqslant r$) is the matrix $\phi(g_i^{-1}gg_j)$ if $g_i^{-1}gg_j \in H$, and the zero matrix otherwise. It is not difficult to check that the map $g \to \psi(g)$ ($g \in G$) is a representation of G; it is called the representation *induced* by the representation ϕ of H, or simply an induced representation (if there is no need to be more explicit). For fixed i, there is a unique element g_j such that $g_i^{-1}gg_j \in H$ (since $g^{-1}g_i$ lies in just one coset g_jH). Similarly, i is uniquely determined by j. Therefore, $\psi(g)$ is a block monomial matrix with blocks of dimension n. In particular, $\psi(G)$ is contained in the matrix wreath product $\phi(H) \wr \mathrm{Sym}_r$.

2.13. Action of a Linear Group on the Algebra of Noncommuting Polynomials. In order to avoid introducing new terminology, from now on to the end of the section, we shall assume that K is an arbitrary field, P say. In fact, almost all the facts introduced below are valid for an arbitrary ring K with identity.

Let $X = \{X_1, \ldots, X_n\}$ be a set of formal symbols. We shall call an expression like $X_{i_1} \ldots X_{i_r}$ a monomial of degree r. Monomials are said to be different if their expressions look different (having regard to the order in which the X_i occur). We define the product of two monomials to be the result of juxtaposing them. This operation is associative, and the empty monomial acts like the identity. Further, let F be the set of formal finite sums of the form $\sum \alpha_i d_i$, where $\alpha_i \in P$ and the d_i are different monomials. The elements of F are called polynomials. We define an addition in F by adding the coefficients of like monomials, and multiplication by elements α of P by multiplying each coefficient by α. Clearly, these operations turn F into an (infinite-dimensional) vector space with the monomials as a basis. Multiplication of monomials extends by distributivity to the whole of F. Under these operations, F becomes an algebra called the *algebra of noncommuting polynomials* over P. The homogeneous polynomials of degree r form a linear subspace, and we shall denote it by $F_{(r)}$. Note that F is also called the free algebra with identity on the basis X.

The group $G = \mathrm{GL}_n(P)$ has a canonical action on the space $F_1 = \{\alpha_1 X_1 + \cdots + \alpha_n X_n\}$ of linear polynomials. We extend this action to the whole of F: for g in G and $f = f(X_1, \ldots, X_n)$, we set $gf = f(gX_1, \ldots, gX_n)$. It is not difficult to verify that this formula defines an isomorphic embedding of G into the automorphism group of F, and in particular it defines an infinite-dimensional representation $\tau: G \to \mathrm{GL}(F)$. Cleary, $\tau(g)F_{(r)} = F_{(r)}$ for all g in G and all $r = 1, 2, \ldots$; a more detailed analysis shows that the restriction τ_r of τ to $F_{(r)}$ is equivalent to the r-th tensor power of τ_1 (see 2.10).

All that has been said applies to an arbitrary subgroup H of G, of course. Elements f of F such that $hf = f$ for all h in H are called invariants of H in F.

2.14. Action of a Linear Group on the Algebra of Polynomials. Let I be the ideal of F generated by the elements $X_{i_1}X_{i_2} - X_{i_2}X_{i_1}$ ($i_1, i_2 \in \{1, \ldots, n\}$). It is clear that the factor-algebra F/I is isomorphic to the usual polynomial algebra S in commuting variables. Moreover, $\tau(g)I = I$. Thus, τ defines an action of the elements g of G on F/I, that is, on S. We denote the resulting (infinite-dimensional) representation by θ. Its restriction to the space $S_{(r)}$ of homogeneous polynomials of degree r is denoted by θ_r. This representation θ_r is called the r-th symmetric power of θ_1.

Of course, we could extend the action of G on $F_{(1)} \cong S_{(1)}$ directly to one on S, as in 2.13. However, the representation of S as F/I is useful in some situations: for example, we can deduce that the r-th tensor power of any representation of degree more than 1 is reducible when $r > 1$.

The following fact is not difficult to establish: if $p = \mathrm{char}\, P = 0$, then every representation $\theta_r(G)$ is irreducible. The same is true when $p > 0$, $r < p$.

For a subgroup H of G, the polynomials f in S fixed by H are called *invariants* of H. Clearly, they form a ring, or more exactly, a subalgebra of S. The problem of describing the ring of invariants for the various subgroups is the main question in invariant theory – an important algebraic discipline lying at the junction of the theory of linear groups and algebraic geometry.

2.15. Action of a Linear Group on the Grassman Algebra. Let J be the ideal of F generated by the elements $X_{i_1}X_{i_2} + X_{i_2}X_{i_1}$ ($i_1, i_2 \in \{1, \ldots, n\}$), and also the elements X_i^2 if P is of characteristic 2. The algebra $F/J = D$ is called the *Grassman algebra* (or *exterior algebra*) of the space $F_{(1)}$.

Since $F_{(1)} \cap J = \{0\}$, $F_{(1)}$ is embedded isomorphically in D. Clearly, $\tau(g)J = J$ for all g in G, so that τ induces a representation γ of G in $\mathrm{GL}(D)$.

It is easy to see that the dimension of the Grassman algebra is finite and equal to 2^n. Let $D_{(r)}$ be the image of $F_{(r)}$ under the natural map $F \to F/J = D$. Clearly, $\gamma(g)D_{(r)} = D_{(r)}$. We denote the restriction of γ to $D_{(r)}$ by γ_r; it is called the r-th exterior power of γ_1 and is usually denoted by Λ^r. Its dimension is the binomial coefficient C_r^n.

The representations $\Lambda^r(G)$, $1 \leqslant r \leqslant n$, are irreducible. So are the representations $\Lambda^r(\mathrm{SL}_n(P))$, which play an important part in the representation theories of algebraic groups and Lie groups. They realise the so-called fundamental representations of $\mathrm{SL}_n(P)$.

We note that the representations τ_1, θ_1, Λ^1 are equivalent. If we fix not $X_1, \ldots, X_n$, but simply $V = F_{(1)}$ itself, the corresponding algebras are denoted by $F[V]$, $S[V]$ and $D[V]$ and called the tensor, symmetric and exterior algebras respectively of the space V.

§ 3. Symplectic, Orthogonal and Unitary Groups

3.1. Sesquilinear Forms. Let V be a space of dimension n over the field P and set $p = \operatorname{char} P$. Let σ be an automorphism of P such that $\sigma^2 = 1$ (the case $\sigma = 1$ is not excluded). A function $f(x, y)(x, y \in V)$ with values in P is said to be a *sesquilinear form* if it is linear in the first argument for fixed values of the second, and semilinear in the second with fixed first. Semilinearity means that the following requirement holds:

$$f(x, y_1 + y_2) = f(x, y_1) + f(x, y_2), \quad f(x, \alpha y) = \alpha^\sigma f(x, y) \; (\alpha \in P).$$

If $\sigma = 1$, f is said to be *bilinear*. If $\sigma \neq 1$, and $f(y, x) = f(x, y)^\sigma$ for all x, y in V, f is said to be *Hermitian*. A bilinear form is *symmetric* if $f(y, x) = f(x, y)$ for all x, y in V, *skewsymmetric* if $f(x, y) = -f(y, x)$, and *alternating* if $f(x, x) = 0$ for all x in V. If $p \neq 2$, a form is skewsymmetric if and only if it is alternating. For $p = 2$, the alternating forms are the ones of most interest to us: the situation is that the study of the isometry group (3.7) of a symmetric form reduces to the alternating case when $p = 2$ (see Dieudonné (1971), Chapter 1, § 10]). To avoid overloading the exposition with preconditions, we shall always reserve the term symmetric form to mean a form over a field of characteristic not 2.

Vectors v, w in V are said to be *orthogonal* if $f(v, w) = 0$. *A* form f is *reflexive* if the relation of orthogonality is symmetric, that is, $f(w, v) = 0$ whenever $f(v, w) = 0$.

3.2. The Gram Matrix. Let $B = \{b_1, \ldots, b_n\}$ be a basis for V. The matrix $\Gamma_f = \{f(b_i, b_j)\}$ is called te *Gram matrix* of f. On change of basis from B to $gb_1, \ldots, gb_n$ for any g in $\mathrm{GL}_n(P)$, Γ_f is replaced by ${}^t g \Gamma_f g^\sigma$. The form f is said to be *non-degenerate* if Γ_f is nonsingular. Two forms are said to be *equivalent* if their Gram matrices relative to suitable bases are the same.

3.3. Classification Problems for Sesquilinear Forms. It is a remarkable fact that every reflexive form is (to within a scalar multiple) symmetric, skew-symmetric or Hermitian.

Nondegenerate alternating forms exist in even dimensions only. Any two nondegenerate alternating forms are equivalent. All Hermitian and symmetric forms have diagonal Gram matrices in suitable bases; bases of this sort are said to be *orthogonal*. An orthogonal basis $e_1, \ldots, e_n$ is said to be *orthonormal* if $f(e_i, e_i) = 1$ for all $i = 1, \ldots, n$.

If P is algebraically closed, there is always an orthonormal basis relative to a nondegenerate symmetric form; thus, nondegenerate symmetric forms of the same dimension are equivalent. For finite P, any two Hermitian forms are equivalent (and there is an orthonormal basis), and symmetric forms comprise two equivalence classes. We note that inequivalent forms can become equivalent when one of them is multiplied by a scalar (that is, $f(x, y)$ is replaced by $tf(x, y)$, $t \in P$); this is another (not very important) possibility for simplifying the classification of forms.

There is a complete classification of symmetric and Hermitian forms over the field of algebraic numbers and over finite fields. For completely arbitrary fields, this problem seems incapable of effective solution.

Classification problems are intimately connected with the discovery of canonical shapes for a form and its Gram matrix. For a skewsymmetric form, a convenient basis is one relative to which the Gram matrix is $\operatorname{diag}\left(\begin{pmatrix} 0 & 1 \\ -1 & 0 \end{pmatrix}, \ldots, \begin{pmatrix} 0 & 1 \\ -1 & 0 \end{pmatrix}\right)$; orthogonal bases for the symmetric and Hermitian cases. In many situations, the most convenient bases are Witt bases (3.5).

3.4. Index of a Sesquilinear Form. Let f be a nondegenerate reflexive form on V. For a subset M of V, we set $M^{\perp} = \{v \in V \mid f(v, M) = 0\}$. If W is a subspace of V, we have $\dim W + \dim W^{\perp} = \dim V$ and $W^{\perp\perp} = W$. A subspace W is said to be *isotropic* if the restriction $f|_W$ is the zero form; *nondegenerate* if $f|_W$ is nondegenerate, and *degenerate* otherwise. A vector v is *isotropic* if $f(v, v) = 0$, and *anisotropic* otherwise.

All the maximal isotropic subspaces have the same dimension, r say. This number is called the *index* of f. In general, the index can take the values $0, 1, \ldots, [n/2]$ (integer part).

For arbitrary P, the index of an alternating form is $n/2$. If P is algebraically closed, the index of a symmetric form is $[n/2]$. If P is finite, the index of a Hermitian form is $[n/2]$. For the field of rational numbers, the index can take any one of the above values.

3.5. Witt Bases. Let W be an isotropic subspace of V of dimension k. There exists a basis $b_1, \ldots, b_k, b_{k+1}, \ldots, b_{n-k}, b_{n-k+1}, \ldots, b_n$ of V such that $b_1, \ldots, b_k \in W$, $b_1, \ldots, b_{n-k} \in W^{\perp}$, $f(b_i, b_j) = 0$ for $k+1 \leqslant i \leqslant n$, $n-k+1 \leqslant j \leqslant n$, and $f(b_i, b_{n-k+j}) = \delta_{ij}$. Such a basis is called a *Witt basis* with respect to W. With respect to a Witt basis, the Gram matrix takes the form

$$\begin{bmatrix} 0 & 0 & E_k \\ 0 & \Psi & 0 \\ \varepsilon E_k & 0 & 0 \end{bmatrix},$$

where Ψ is the Gram matrix of the restriction of f to the subspace $\langle b_{k+1}, \ldots, b_{n-k} \rangle$ with respect to this basis, $\varepsilon = -1$ if f is skewsymmetric, and $\varepsilon = 1$ otherwise. If $\varepsilon = 1$, the elements $b_{k+1}, \ldots, b_{n-k}$ can be chosen so that Ψ is diagonal (except for the case of an alternating form in characteristic 2). If $n = 2k$, Ψ does not appear.

3.6. Symplectic, Orthogonal and Unitary Groups. Let f be a sesquilinear form on V. An element g of $\mathrm{GL}(V)$ is said to be an *isometry* relative to f if $f(gx, gy) = f(x, y)$ for all x, y in V. The isometries form a group, which we denote by $I(f, V)$. Choice of a basis in V produces a realisation of $I(f, V)$ in terms of matrices; it is appropriate to denote this realisation by $I(\Gamma_f, P)$, where Γ_f is the Gram matrix. It is easy to see that $I(\Gamma_f, P) = \{g \in \mathrm{GL}_n(P) | {}^t g \Gamma_f g^\sigma = \Gamma_f\}$, where t means transpose.

Let f be a nondegenerate form. If f is Hermitian, the isometry group is said to be *unitary* and is denoted by $U(f, V)$. If f is alternating, the isometry group is symplectic and denoted by $\mathrm{Sp}(f, V)$. In the case of a symmetric form with char $P \neq 2$, we use the term "*orthogonal*" group, notation $O(f, V)$. Note that orthogonal groups in characteristic 2 are defined rather differently (see 3.9, below). If Γ_f is the Gram matrix of f relative to some basis, we denote the corresponding matrix groups by $U(\Gamma_f, P)$, $\mathrm{Sp}(\Gamma_f, P)$ or $O(\Gamma_f, P)$. Since the Gram matrix determines the form uniquely, we can omit the letter f. This is not just to simplify expressions: given an arbitrary nonsingular matrix Γ which is symmetric, skewsymmetric or Hermitian (that is, ${}^t\Gamma = \Gamma^\sigma$; in characteristic 2, instead of skewsymmetric matrices we have to take symmetric matrices with zeros along the diagonal), we can define the group $I(\Gamma, P) = \{g \in \mathrm{GL}_n(P) | {}^t g \Gamma g^\sigma = \Gamma\}$, which is identical with the group $I(\Gamma_f, P)$ associated with the corresponding form f. This means that the symplectic, orthogonal and unitary groups can be given in purely matrix form, without reference to a vector space. However, it is significantly more difficult to work with these groups using only the apparatus of matrices than it is with the apparatus of linear algebra.

These groups are usually referred to as *classical groups* ($\mathrm{GL}_n(P)$ and $\mathrm{SL}_n(P)$ are sometimes included under this heading). The related groups, namely the *special unitary group* $\mathrm{SU}(\Gamma, P)$ and the *special orthogonal group* $\mathrm{SO}(\Gamma, P)$ are defined by taking only matrices of determinant 1 (like $\mathrm{SL}_n(P)$ in $\mathrm{GL}_n(P)$). Note that all matrices in $\mathrm{Sp}(\Gamma, P)$ are of determinant 1. For fixed n, the groups $\mathrm{Sp}(\Gamma, P)$ are all similar; thus, the Γ is often omitted and we write $\mathrm{Sp}_n(P)$ (or $\mathrm{Sp}_{2m}(P)$ as n is even). Not infrequently one needs to consider the central factor-groups of classical groups. Their names usually involve the word "projective" (as in projective special linear group, projective symplectic group, projective unitary group *etc*), and the notation has the letter P at the beginning: $\mathrm{PSL}_n(P)$, $\mathrm{PSp}_{2m}(P)$, $\mathrm{PSU}_n(P)$. The word "general" is usually omitted from the name of $\mathrm{PGL}_n(P)$. When the ground field is finite, $|P|$ is often written instead of P; for example, $\mathrm{PGL}_2(5)$ is the projective linear group of degree 2 over the field of five elements. See Dieudonné (1971) for formulae giving the orders of classical groups over finite fields; these groups are often called the finite classical groups.

3.7. Examples. The group $O(E_n, P)$ is called the *group of orthogonal matrices*, and is usually denoted by $O_n(P)$. Set $\Gamma^k = \mathrm{diag}(E_k, -E_{n-k})$. For $0 < k < n$, $O(\Gamma^k, \mathbb{R})$ is sometimes said to be *pseudo-orthogonal* (in particular, this happens in Lie group theory). Every nondegenerate symmetric bilinear form over $\mathbb{R}$ is equivalent to a form with Gram matrix Γ^k for some k, $0 \leq k \leq n$. When k is replaced by $n - k$, $O(\Gamma^k, P)$ is replaced by a group similar to it. It follows from this that every orthogonal group $O(\Gamma_f, \mathbb{R})$ is similar to $O(\Gamma^k, \mathbb{R})$ for some k with $0 \leqslant k \leqslant [n/2]$.

For $n = 4$, $k = 1$, $O(\Gamma^1, \mathbb{R})$ is called the *Lorenz group*. It plays an important part in special relativity theory.

The group $U(E_n, P)$ is called the *group of unitary matrices* and is usually denoted by $U_n(P)$. If $P = \mathbb{C}$ and σ means complex conjugation, the groups $U(\Gamma^k, \mathbb{C})$ $(0 < k < n)$ are frequently called *pseudo-unitary*. The unitary groups $U(f, \mathbb{C})$ are determined up to similarity by the index of f; every $U(f, C)$ is conjugate in $\mathrm{GL}_n(\mathbb{C})$ to one of the groups $U(\Gamma^k, \mathbb{C})$ with $0 \leqslant k \leqslant [n/2]$.

3.8. Sesquilinear Forms and Involutions. Let End V be the ring of linear maps on a space V; it is realised as the ring of $n \times n$ matrices with respect to some basis of V, where $n = \dim V$. Let f be a nondegenerate reflexive form on V. For $A \in \operatorname{End} V$, there is a unique element $A^* \in \operatorname{End} V$ such that $f(Ax, y) = f(x, A^*y)$ for all x, y in V. Moreover, the map $A \to A^*$ has the following properties: $A^{**} = A$, $(AB)^* = B^*A^*$, $(A + B)^* = A^* + B^*$ (for all A, $B \in \operatorname{End} V$), so that $A \to A^*$ is an *involutory antiautomorphism* (or involution for short) of End V. Much deeper is the fact that, for every involution θ of End V, there is a reflexive form f on V such that $A^* = \theta(A)$ for all A in End V.

Note that the relation $f(gx, gy) = f(x, y)$ for all x, y in V is equivalent to the equation $g^*g = E$. It follows that the unitary, orthogonal and symplectic groups can be specified as the subgroup consisting of the elements g of $\mathrm{GL}(V)$ such that $g^*g = E$. This fact is very useful in many situations.

3.9. Quadratic Forms. A function $q(x)$, $x \in V$, is said to be a *quadratic form* on V if $q(\alpha x) = \alpha^2 q(x)$ for all $x \in V$, $\alpha \in P$, and $f(x, y) = q(x + y) - q(x) - q(y)$, for x and y in V, is a bilinear form on V. Clearly, the group $I(q, V) = \{g \in \mathrm{GL}(V) | q(gx) = q(x) \text{ for all } x \in V\}$ is contained in $I(f, V)$.

The form f is symmetric; if char $P = 2$, then $f(x, x) = 0$, that is, f is alternating. If char $P \neq 2$, the quadratic form q can be recovered from f, because $f(x, x) = 2q(x)$; thus $I(q, V) = I(f, V)$ in this case.

The theory of quadratic forms over a field of characteristic 2 has a whole range of peculiarities. We shall consider the simpler situation when f is nondegenerate. In this case, $I(q, V)$ is called the *orthogonal group* of the form q and is denoted by $O(q, V)$. It turns out that $O(q, V)$ is a proper subgroup of $\operatorname{Sp}(f, V)$. It is of great importance when the ground field is finite; in this case, the commutator subgroup is a simple group when $\dim V > 4$.

The case where f is degenerate is of less interest. If the ground field is perfect, study of the group $I(q, V)$ reduces to that of the orthogonal or symplectic group of smaller dimension. If P is not perfect, some new curiosities arise. However, the case of a non-perfect field of characteristic 2 seems to be quite special, and we shall not be considering it. It is covered in fair detail in Dieudonné (1971).

A nonzero vector v in V is said to be *singular* if $q(v) = 0$; a nonzero subspace W of V is *singular* with respect to the form q if it is isotropic relative to the associated form f and every nonzero vector w in W is singular. When char $P \neq 2$, it is clear that the singular vectors (subspaces) are just the isotropic vectors (subspaces). However, these concepts are quite different in characteristic 2.

3.10. Witt's Theorem. Let f be a nondegenerate Hermitian, symmetric or alternating form. Let W_1 and W_2 be subspaces of V, and ϕ: $W_1 \to W_2$ an isomorphism such that $f(\phi(x), \phi(y)) = f(x, y)$ for all $x, y \in W_1$. Witt's Theorem asserts that there is an element g in $I(f, V)$ such that $\phi(x) = g(x)$ for all x in W_1. In other words, every isometric linear map from W_1 onto W_2 can be extended to an isometry of V. In essence, this is the fundamental theorem for classical groups.

There is an analogue of Witt's Theorem for orthogonal groups in characteristic 2. Let q be a quadratic form on V with nondegenerate form f, and let W_1, W_2 be subspaces of V. If ϕ: $W_1 \to W_2$ is an isomorphism preserving q, that is, $q(x) = q(\phi(x))$ for all x in W_1, there exists an element g of $O(q, V)$ such that $g(x) = \phi(x)$ for all x in W_1 (Bourbaki (1958); Chapter IX, §4).

We note some important special cases of Witt's Theorem. (A) The group $I(f, V)$ acts transitively on the isotropic subspaces of given dimension (and also on vectors x in V of equal "length" $f(x, x)$). A subspace W of V is said to be *hyperbolic* if it is nondegenerate and contains an isotropic subspace of dimension dim $V/2$ (by (3.5), this is equivalent to the requirement that W be a sum of two isotropic subspaces and $W \cap W^{\perp} = \{0\}$). (B) The group $I(f, V)$ acts transitively on the hyperbolic subspaces of every fixed dimension.

3.11. The Stability Group of an Isotropic Space. Let W be an isotropic space and B a Witt basis relative to W (see 3.4). It is useful to have a description of the group $I_W = \{g \in I(f, V) | gW = W\}$. It is clear that it also leaves $W^{\perp}$ invariant, and $W \subset W^{\perp}$ in this case. Set $k = \dim W$, $U = \langle b_{k+1}, \ldots, b_{n-k} \rangle$ and $f_U = f|_U$. We consider two subgroups of the group of block-diagonal matrices $D_{[\lambda]}(P)$, where $\lambda = \{k, n-2k, k\}$. Namely, set $K = \mathrm{diag}(A, E_{n-2k}, ({}^tA^{\sigma})^{-1})$, $L = \mathrm{diag}(E_k, C, E_k)$, where A runs over $\mathrm{GL}_k(P)$ and C runs over $I(f_U, U)$. The groups K and L permute elementwise, and $I_W = KLI^0$ (semidirect product), where I_0 is the set of elements g of I_W acting like the identity on W and on the factors $W^{\perp}/W$, $V/W^{\perp}$. In other words, I^0 is the kernel of the natural homomorphism $I_W \to D_{[\lambda]}(P)$, the image of which is KL. There is some benefit in giving an explicit description of the matrices in I^0. We write

$$h = \begin{pmatrix} E_k & q & s \\ 0 & E_{n-2k} & r \\ 0 & 0 & E_k \end{pmatrix},$$

where q, r, s are matrices of the relevant dimensions; we shall find relations between them following from the fact that h lies in $I(f, V)$. The matrix Γ_f is described in 3.5. In the notation of 3.4, the condition ${}^th\Gamma_f h^{\sigma} = \Gamma_f$ is equivalent to the pair of relations $\varepsilon q^{\sigma} + {}^tr\Psi = 0$, $\varepsilon s^{\sigma} + {}^ts = -{}^tr\Psi r^{\sigma}$ (so that, if $q = 0$, we have $r = 0$ and ${}^ts = -\varepsilon s^{\sigma}$). We note that the matrix $s = -\varepsilon {}^tr^{\sigma}\Psi r/2$ satisfies these conditions if char $P \neq 2$. It follows that

$$h = \begin{bmatrix} E_k & -\varepsilon^t r^\sigma \Psi^\sigma & -\varepsilon^t r^\sigma \Psi r/2 \\ 0 & E_{n-2k} & r \\ 0 & 0 & E_k \end{bmatrix} \cdot \begin{bmatrix} E_k & 0 & s_1 \\ 0 & E_{n-2k} & 0 \\ 0 & 0 & E_k \end{bmatrix},$$

where ${}^t s_1 = -\varepsilon s_1^\sigma$. In general, the matrices satisfying $q = 0$, ${}^t s = -\varepsilon s^\sigma$ form an abelian normal subgroup of I^0 (and even of I_W).

3.12. Spinor Groups and Clifford Algebras. Let q be a quadratic form on V (3.9). We fix some basis $X = \{X_1, \ldots, X_n\}$ of V and consider the ideal L of the algebra F of noncommuting polynomials (2.13) of $X_1, \ldots, X_n$ generated by the elements $x^2 - q(x)\cdot 1$, where 1 is the identity of F and x runs over the whole of V. The factor-algebra F/L is the *Clifford algebra* of q. A different choice of basis gives rise to an isomorphic algebra, so we can denote the Clifford algebra by $C(V)$. It is 2-*graded*: $C(V) = C^+ + C^-$ (this is a direct sum of subspaces such that $C^+C^+ \subset C^+$, $C^-C^- \subset C^+$, $C^+C^- \subset C^-$, $C^-C^+ \subset C^-$). In particular, C^+ is a subalgebra. The components C^+ and C^- of the grading are constructed as follows. Let F^+ and F^- be the sums of the subspaces F_k (of polynomials of degree k) of even and odd degrees respectively. Then, the decomposition $F = F^+ + F^-$ gives a 2-grading, and since the elements $x^2 - q(x)\cdot 1$ lie in F^+, this 2-grading is inherited by the factor-algebra $F/L = C(V)$; that is, we can take C^+ and C^- to be the images of F^+ and F^- under the homomorphism $F \to F/L$. The algebra C^+ is simple and has centre P; its dimension over P is 2^{2n}.

If char $P \neq 2$, a convenient choice for X is an orthogonal basis relative to the bilinear form f associated with q (see 3.9). The relation $x^2 - q(x)\cdot 1 = 0$ is satisfied in $C(V)$ if (and only if)

$$(*) \quad x_i^2 = q(x_i)\cdot 1, \quad x_i x_j = -x_j x_i \quad (i, j = 1, \ldots, n).$$

In other words, the Clifford algebra can be given in terms of generators subject to relations (*).

Since $V \cap L = \{0\}$, V can be considered to be a subspace of $C(V)$. The Clifford group Cl_V is the group of invertible elements g in $C(V)$ such that $gVg^{-1} = V$.

There is a unique antiautomorphism β of $C(V)$ acting as the identity on V. We have $\beta^2 = 1$. The map $a \to \beta(a)a$, $a \in C(V)$, is a homomorphism of the Clifford group into the centre P of C^+. The scalar $\beta(a)a$ in P is called the *spinor norm* of the element a of $C(V)$. The elements a of $C(V)$ with spinor norm 1 form the *spinor group* Spin_V.

The map $v \to ava^{-1}$ ($v \in V$, $a \in Cl_V$) defines a homomorphism of the Clifford group into $O(q, V)$ whose kernel lies in the centre. The image $\Omega(q, V)$ of the spinor group under this homomorphism is called the reduced orthogonal group. If this group contains a unipotent element, then it is generated by unipotent elements and is equal to its commutator subgroup. See Dieudonné (1971) for details.

3.13. Groups of Unitary and Orthogonal Matrices. These groups were introduced in 3.7. It follows from 3.8 that the group $U_n(\mathbb{C})$ consists of all matrices such that ${}^t\overline{U}U = E_n$, where the bar denotes complex conjugation. Such matrices are said to be *unitary*. Similarly, we have $O_n(P) = \{g \in \mathrm{GL}_n(P) | {}^tg = g^{-1}\}$. Matrices with this property are said to be orthogonal (more especially when $P = \mathbb{R}$).

Suppose that $U_n(\mathbb{C})$ acts in the natural way on a vector space V with basis $b_1, \ldots, b_n$. As we have already seen, it leaves invariant a form h with Gram matrix E_n. If $x = x_1b_1 + \cdots + x_nb_n$ and $y = y_1b_1 + \cdots + y_nb_n$ are any two vectors in V, then $h(x, y) = x_1\bar{y}_1 + \cdots + x_n\bar{y}_n$. In particular, $U_n(\mathbb{C})$ preserves the length $d(x)$ of vectors x in V, defined by $d(x) = (x_1\bar{x}_1 + \cdots + x_n\bar{x}_n)^{1/2}$; that is, $d(ux) = d(x)$ for u in $U_n(\mathbb{C})$. Similarly, $O_n(\mathbb{R})$ preserves the length $d(x) = (x_1^2 + \cdots + x_n^2)^{1/2}$ of vectors x in Euclidean space. This is one of the main reasons for the importance of these groups in physics.

We record some important properties of the group of unitary matrices.

(A) The group $U_n(\mathbb{C})$ is compact and every compact (in particular, every finite) subgroup of $\mathrm{GL}_n(\mathbb{C})$ is similar to a subgroup of $U_n(\mathbb{C})$.

The topology here is the Euclidean topology defined by the metric $\mu(x, y) = d(x - y)$, for $x, y \in M_n(\mathbb{C})$, and the basis chosen for $M_n(\mathbb{C})$ is that consisting of the matrix units.

(B) Every eigenvalue of a unitary matrix is of modulus 1.

(C) The length of the vector formed by any row or column of a unitary matrix is 1 (the origin of the term "unitary matrix" is connected with this fact).

(D) Every abelian group of unitary matrices is similar to a subgroup of the group $D_n(\mathbb{C})$ of diagonal matrices; in particular, every unitary matrix is diagonalisable.

Properties (A), (B), (C) also hold for the groups of orthogonal matrices (with $\mathbb{C}$ replaced by $\mathbb{R}$). Property (D) does not go over to $O_n(\mathbb{R})$.

3.14. Remarks. We draw the attention of the reader intending to consult other sources to the fact that terminology and notation associated with sesquilinear forms and their isometry groups can be rather different in the works of other authors. For example, unitary group sometimes means isometry group of a nondegenerate sesquilinear ε-Hermitian form $f(x, y)$, where $\varepsilon = 1$ or -1 (that is, a form such that $f(y, x) = \varepsilon f(x, y)^\sigma$), so that orthogonal and symplectic groups are sometimes viewed as special cases of unitary groups. Instead of our term "isotropic subspace", one sometimes uses "completely isotropic", *etc.*

Familiarity with the classical groups can be broadened by consulting the books of Artin (1957), Dieudonné (1971) and Chapter IX of Bourbaki (1958).

§4. Linear Groups over Fields with Valuation

The classical examples of fields with valuation are the complex and real fields, with absolute value as valuation. The valuation on a field extends naturally to vector spaces over it, and in particular to spaces of matrices, where it defines metrics and topologies. The techniques associated with these concepts play an important part not only in the theory of linear groups, but also in very many other investigations. In particular, linear Lie groups, especially semisimple Lie groups and their representations, have numerous applications in theoretical physics.

The theory of p-adic Lie groups is developed in parallel with these theories. The main applications are in number theory and algebraic geometry. In recent times there has been a significant expansion of the application of local analytic methods in the theory of linear groups, that is, methods involving the embedding of linear groups in groups over local fields (see 2.3.2). We note that local analysis has long played a major role in the representation theory of finite groups and thereby in the workings of finite linear groups.

4.1. Euclidean Vector Spaces. In sections 4.1–4.5, we shall assume that the ground field P is $\mathbb{C}$ or $\mathbb{R}$. Let V be a vector space over P with a fixed basis $b_1, \ldots, b_n$. We shall denote the usual modulus of a complex number α by $\|\alpha\|$. Let $v = \alpha_1 b_1 + \cdots + \alpha_n b_n$ be any element of V. We define the norm $\|v\|$ to be the length of v, that is, $\|v\| = (\sum \|\alpha_i\|)^{1/2}$, while the distance $\mu(a, b)$ between points a and b with coordinates $(\alpha_1, \ldots, \alpha_n)$ and $(\beta_1, \ldots, \beta_n)$ is given by the formula $d(a, b) = (\sum \|\alpha_i - \beta_i\|)^{1/2}$. The space V with this distance function (metric) is said to be an *Euclidean space*, and the topology induced by this metric the *Euclidean topology*. Our definition of Euclidean space presupposes a choice of basis; however, it is not hard to check that the topology is independent of the basis, and that the corresponding distance functions are equivalent. What this means is: if μ_1, μ_2 are two such functions, there exist positive constants c_1 and c_2 in $\mathbb{R}$ such that $c_1 \mu_2(a, b) \leqslant \mu_1(a, b) \leqslant c_2 \mu_2(a, b)$ for all a and b in V.

In the same way, if we choose as basis for $M_n(P)$ the n^2 matrix units e_{ij} $(i, j = 1, \ldots, n)$, we introduce a Euclidean space structure on $M_n(P)$. The choice of a basis in V defines compatible Euclidean space structures on V and on $M_n(P)$. We recall (3.13) that the map $v \to \|v\|^2$ is a symmetric quadratic form on V when $P = \mathbb{R}$, and that $O_n(\mathbb{R})$ is precisely the group of matrices whose action on V preserves the norm. A similar thing can be done for $U_n(\mathbb{C})$. We note also that right (and left) multiplication by elements of $U_n(\mathbb{C})$ preserves the distance function on $M_n(\mathbb{C})$ defined above (and similarly for $P = \mathbb{R}$).

4.2. Linear Groups Closed in the Euclidean Topology. It is natural to discuss the question of what the linear groups over $\mathbb{R}$ or $\mathbb{C}$ that are closed in the Euclidean topology look like. Since linear groups over $\mathbb{R}$ that are closed in the real Euclidean topology are closed in the Euclidean topology on $M_n(\mathbb{C})$, it is enough to speak about closed subgroups of $\mathrm{GL}_n(\mathbb{C})$. Note that $\mathrm{GL}_n(\mathbb{C})$ itself is not closed, but is isomorphic to a closed subgroup of $\mathrm{GL}_{n+1}(\mathbb{C})$ (simply associate each matrix g in $\mathrm{GL}_n(\mathbb{C})$ with the matrix $\mathrm{diag}(g, \det(g^{-1}))$ in $\mathrm{GL}_{n+1}(\mathbb{C})$). Most of the groups that we have been considering in preceding sections are closed. The class of closed groups is very extensive. The connected groups and the discrete groups form the most important subclasses. The typical example of a discrete linear group is $\mathrm{SL}_n(\mathbb{Z})$.

Every closed linear group over $\mathbb{R}$ or $\mathbb{C}$ is a Lie group. (The converse is false: linear Lie groups can fail to be closed.) Linear Lie groups have been very intensively studied in Lie group theory; see for example the book Freudenthal and de Vries (1969). The most complete structural results in the theory have to do with connected Lie groups. Examples are the well-known groups $\mathrm{SL}_n(\mathbb{C})$, $\mathrm{Sp}_{2m}(\mathbb{C})$, $U_n(\mathbb{C})$. Discrete subgroups of connected Lie groups have received particular attention, especially subgroups of $\mathrm{SL}_n(\mathbb{C})$. We cannot, however, immerse ourselves in the deep waters of Lie group theory. Our aim here is to cover a number of facts that explain the connection of Lie group theory with linear groups.

4.3. Connected Linear Lie Groups. A connected Lie group is said to be *simple* (*semisimple*, respectively) if it contains no proper connected normal subgroups (no proper connected abelian normal subgroups, respectively) other than 1.

Every connected Lie group contains a unique maximal connected soluble normal subgroup $R(G)$, and a maximal connected semisimple subgroup M defined uniquely up to conjugacy in G. The group $M \cap R(G)$ is discrete, lies in the centre of M, and the group G has the decomposition $G = M \cdot R(G)$, called the Levi-Mal'tsev decomposition. If G is linear, then $M \cap R(G)$ is finite. Every semisimple Lie group is a product of elementwise-permuting simple Lie groups. The connected simple Lie groups have been classified fully. In particular, every simple Lie group with trivial centre is isomorphic to a connected component of the automorphism group of some Lie algebra over $\mathbb{R}$ or $\mathbb{C}$.

Every connected soluble linear Lie group is similar over $\mathbb{C}$ to a subgroup of the group of upper-triangular matrices. A connected linear Lie group G is completely reducible if and only if $R(G)$ does not contain unipotent matrices. If G is reducible, its irreducible components are also Lie groups. An irreducible connected linear Lie group is the Kronecker product of simple linear Lie groups and a Lie subgroup of $\mathrm{GL}_1(\mathbb{C})$. Therefore, the structure of G more-or-less reduces to a description of the simple irreducible connected linear Lie groups.

Simple Lie groups are known up to (analytic) isomorphism. Moreover, every simple linear Lie group has a covering group, that is, a simple linear Lie group such that every simple linear Lie group with the same central factor-group (as the given group) is a homomorphic image of the covering group. In this way, the description of the simple linear Lie groups becomes a problem in the representation theory of the covering simple linear groups.

Their analytic representations are described in parallel with the representations of the corresponding Lie algebras. The theory of the irreducible representations of simple Lie algebras has been very thoroughly investigated. In particular, formulae are known giving the degrees of the irreducible representations of all simple Lie algebras, and thus of their Lie groups.

One of the most remarkable facts in the theory of Lie groups is the following result about compact subgroups.

Theorem. *Let G be a connected linear Lie group. Every compact subgroup of it is contained in a maximal compact subgroup. All maximal compact subgroups are connected and are conjugate in G.*

A linear Lie group is said to be *reductive* if it contains no unipotent normal subgroups. The maximal abelian reductive subgroups in connected Lie groups are called the Cartan subgroups. Every Cartan subgroup is closed and connected. A connected (linear) Lie group has only finitely many conjugacy classes of Cartan subgroups. In a compact connected Lie group, all Cartan subgroup are conjugate.

There are other interesting results on closed connected linear groups coming from Lie group theory. However, it seems reasonable to restrict attention to the above examples and to refer the interested reader to the relevant texts (Freundenthal and de Vries (1969), Zhelobenko and Shtern (1983) etc).

Discrete subgroups of connected groups occupy an important place in Lie group theory. Principally, the theory describes the connection between the structure of a discrete subgroup Γ and properties of the factor-space G/Γ. The main interest for us is the fact that Γ is almost always finitely generated, and in many important situations it is finitely related. We could not avoid giving many special definitions if we wanted to state these results properly. Thus, we refer the reader to Ragunathan's monograph (1972). However, we shall return later to the finite presentation problem for linear groups (section 2.3.4).

4.4. Fields with Valuation. We recall that a *valuation* on a field P is a map $v: P \to \mathbb{R}$ with the following properties:

$$\text{(a)}\ \ v(0) = 0 \text{ and } v(\alpha) > 0 \text{ for } \alpha \neq 0\ (\alpha \in P),$$

$$\text{(b)}\ \ v(\alpha + \beta) \leqslant v(\alpha) + v(\beta)\ (\alpha, \beta \in P);$$

$$\text{(c)}\ \ v(\alpha\beta) = v(\alpha)v(\beta).$$

The valuation is *trivial* if $v(\alpha) = 1$ for every nonzero element α of P. A field P is said to be a field with valuation if some nontrivial valuation is given on it.

A field P with valuation is said to be *complete* if every fundamental sequence (= Cauchy sequence) in it converges. Every field with valuation has a completion, that is, is contained as a dense subset in a complete field with valuation extending that on P. In the main, we shall be interested in complete fields.

Every valuation on a field P can be extended to finite (and therefore to algebraic) extensions of P; if P is complete, the extension of the valuation is unique.

A valuation v is said to be *non-Archimedean* if the following condition is satisfied:

$$\text{(b}'\text{)}\ \ v(\alpha + \beta) \leqslant \max(v(\alpha), v(\beta)) \quad \text{for all } \alpha, \beta \in P.$$

Otherwise it is *Archimedean.* Non-Archimedean valuations remain non-Archimedean on completion. Every valuation on a field of nonzero characteristic is non-Archimedean.

Let p be a rational prime. There exists a nontrivial non-Archimedean valuation v_p on $\mathbb{Q}$ such that $v_p(q) = 1$ for every prime $q \neq p$. For fixed p, such a valuation is unique in the sense that, for every other valuation v_1 with the property, we have $v_p(\alpha) = v_1(\alpha)^t$ for some t in $\mathbb{R}$ and all α in P. It is known as the p-adic valuation and the completion $\mathbb{Q}_p$ of $\mathbb{Q}$ with respect to it is called the field of p-adic numbers. Every non-Archimedean valuation on $\mathbb{Q}$ is the p-adic valuation for some p. All Archimedean valuations on $\mathbb{Q}$ are of the form $\|\alpha\|^t$, where $0 < t \leqslant 1$ and $\| \;\|$ denotes absolute value.

Every nondiscrete locally compact field is $\mathbb{R}$, $\mathbb{C}$, a finite extension of $\mathbb{Q}_p$ for some p, or the field of formal power series in a single variable with finite field of constants (see van der Waerden (1935), Chapter VI, 9.3], for example). All these fields are complete fields with valuation and their topologies are induced by the valuations.

Every finitely generated field can be embedded in a nondiscrete locally compact field; moreover, for every element t other than a root of 1, there is an embedding such that the valuation of t is different from 1.

Let v be a non-Archimedean valuation on a field P; the set of elements v of P such that $v(\alpha) \leqslant 1$ clearly forms a subring P_v; it is called the *valuation ring* of P with respect to v. Then, elements α of P with $v(\alpha) < 1$ form an ideal of P called the *valuation ideal*, denoted here by I_v. The factor-ring P_v/I_v is a field called the *field of residues*. Two non-Archimedean valuations define one and the same topology on a field P if and only if their valuation rings coincide.

4.5. Linear Groups over Locally Compact (Non-discrete) Fields. In this subsection we shall assume that P is a locally compact field with compatible non-Archimedean valuation (see 4.4). We give $M_n(P)$ the topology of the Cartesian product of n^2 copies of P. The problem of the structure of the subgroups closed in this topology arises naturally. First of all we note that such groups are analytic groups over P. The definition of such groups is fully analogous to that of analytic Lie groups (see Serre (1965), Chapter IV). The theory of analytic groups over non-Archimedean fields was essentially created by Lazard (1965). Instead of "analytic group", Bourbaki (1971–72, 1968, 1975) applies the term "Lie group over P", and develops a uniform theory of Lie groups over arbitrary locally compact fields. The chief object there is the correspondence between analytic groups and Lie algebras and an investigation of the properties of this correspondence.

If char $P = 0$ then P is a finite extension of $\mathbb{Q}_p$, the field of p-adic number. For fixed p one often uses the term p-adic analytic groups. Bass, Milnor and Serre (1967) used the theory of p-adic analytic groups for proving the following surprising theorem.

Theorem A. *Let G be a semisimple algebraic group defined over $\mathbb{Q}$ and Γ an arithmetic subgroup of G. Suppose that $\mathbb{Q}$-rank of G (see* 1.6.11) ≥ 2. *Then every representation of Γ extends to a $\mathbb{Q}$-defined polynomial representation of G.*

It follows that every representation of Γ is completely reducible. In particular, all this is true for $\mathrm{SL}(n, \mathbb{Z})$ $(n > 2)$ and $\mathrm{Sp}(2n, \mathbb{Z})$.

Compact analytical groups are most important for applications to linear groups. Lazard (1965) discovered a group-theoretical characterization of compact p-adic analytical groups; this result was essentially refined by Lubotzky and Mann (1987). We recall first that a *pro-p-group* is defined to be the inverse limit of finite p-groups. A group is called to be *of rank r* if each subgroup of it can be generated by r elements.

Theorem B (Lazard, Lubotzky and Mann). *Every compact p-adic analytical group contains an open subgroup of finite index which is a pro-p-group of finite rank.*

A good exposition of the technique of p-adic analytical groups with applications to group theory may be found in the book of Dixon, du Sautoy, Mann and Segal (1991).

There are a number of significant results for groups that are groups of P-points of algebraic linear groups (see 6.6); these results are obtained using the methods of algebraic group theory.

The following theorem about the general linear groups has important applications: Let $A = P_v$ be the valuation ring of a field P (see 4.4). Then $\mathrm{GL}_n(A)$ is compact. Moreover, every compact subgroup H of $\mathrm{GL}_n(P)$ is conjugate in $\mathrm{GL}_n(P)$ to a subgroup of $\mathrm{GL}_n(A)$. In particular, the maximal compact subgroups of $\mathrm{GL}_n(P)$ are conjugate. Note that the maximal compact subgroups of $\mathrm{SL}_n(P)$ are not necessarily conjugate (though they are conjugate in $\mathrm{GL}_n(P)$).

§ 5. The Zariski Topology

The Zariski topology is important in linear group theory, and it forms the basis of the theory of algebraic linear groups (see § 6). Moreover, the Zariski topology is a useful tool for studying infinite linear groups at an elementary level. Zariski topology techniques have been applied with success to the solution of problems arising in very diverse situations. This is explained in the first place because the Zariski topology gives rise to a well-defined concept of dimension for linear groups, and one can argue by induction on the dimension (see 2.3.3).

5.1. Definition of the Zariski Topology. Let P be any infinite field and V a vector space of dimension k over P. We define the *Zariski topology* by declaring that a set is closed if it is the solution set of a system of polynomial equations over V. In more detail, we realise V as the set k-vectors $(\alpha_1, \dots, \alpha_k)$, and set $S = P[x_1, \dots, x_k]$ for the set of polynomials in k variables with coefficients in P. A subset M of V is said to be Zariski-closed if there is a subset T of S such that $M = \{(\alpha_1, \dots, \alpha_k) | f(\alpha_1, \dots, \alpha_k) = 0 \text{ for all } f \in T\}$.

Proposition. *The intersection of any number of closed sets is closed*; *moreover, V satisfies the minimum condition for closed subsets.*

The second assertion here is another form of the celebrated Hilbert Basis Theorem (every ideal in S has a finite basis).

It is easy to see that change of basis in V leads to the same topology. Clearly, the union of finitely many closed sets is closed; V and the empty set are closed. Every point $(\alpha_1, \dots \alpha_k)$ forms a closed set, since it is the solution set of the equations $x_1 = \alpha_1, \ldots, x_k = \alpha_k$. Consequently, finite sets are closed. All this proves that the axioms for a topology are satisfied. However, we must warn the reader that Hausdorff's separation axiom does not hold. This is completely obvious for $k = 1$, since every closed set is finite. A peculiarity of the Zariski topology is that two nonempty open sets have nonempty intersection. This may cause difficulties for those who habitually work in separable topologies.

5.2. Zariski Dimension. By a *connected set* in V we mean a closed set not expressible as the union of two proper closed subsets. (We are departing here from the standard terminology: however, our reluctance to go into details forces us to push on.) It is a remarkable fact that every sequence of pairwise different connected sets contained one in the other has not more than k terms. This allows us to define the *dimension of a closed set* to be the greatest length of such a sequence of connected sets contained in it. We note that V is a connected set, and its dimension in the sense just defined is k. This avoids confusion when we denote the dimension of a closed subset W of V by dim W. If necessary, we shall employ the term "Zariski dimension".

5.3. Continuity of Operations. The addition of vectors in V is continuous in each variable. In addition, if pairs of elements of V are viewed as elements of $V \oplus V$, we see that addition is continuous in both variables. This does not contradict the known theorem stating that the underlying space of a topological group H is Hausdorff if multiplication in it is continuous in both variables; the fact is that we are equipping the set of pairs $(v, w) \in V \times V$ with the Zariski topology on $V \times V$, which is significantly stronger than the usual product topology (consider the case $k = 1$ as an illustration). Suppose that $k = n^2$, so that V can be identified with $M_n(P)$. Then addition and multiplication of matrices is continuous in each of the variables (and in both variables in the above sense).

5.4. Closed Subgroups. Let G be any subgroup of $\mathrm{GL}_n(P)$. We give G the topology induced from the Zariski topology on $M_n(P)$; that is, the closed sets in G are intersections of G with closed subsets of $M_n(P)$. We shall look at certain properties of G with respect to this topology. Note that the most important case is that where $G = \mathrm{GL}_n(P)$, though this is not essential.

We consider some typical examples of subgroups of G closed in the Zariski topology.

(a) Let M be any subset of $\mathrm{GL}_n(P)$, and $T = C_G(M)$ the *centraliser* of M in G, that is, the set of elements of G commuting with each element of M. Then T is a closed subgroup.

(b) Let M be a closed subset of G. Then the *normaliser* $N_G(M) = \{g \in G \mid gMg^{-1} = M\}$ is a closed subgroup. This is also true when M is a closed subset of $M_n(P)$, in particular when it is a subspace.

(c) As usual, the *closure* $\overline{M}$ of a subset M of G is defined to be the smallest closed set containing M. The closure $\overline{H}$ of every abstract subgroup H of G is a closed subgroup.

(d) Let G be a group of matrices with the natural action on a space V. If $v \in V$, the stabilizer $\mathrm{Stab}_G(v) = \{g \in G | gv = v\}$ of an element v in G is a closed subgroup. If M is a subset of V closed in the Zariski topology on V, then $\mathrm{Stab}_G(M) = \{g \in G | gM = M\}$ is a closed subgroup of G.

(e) Let W be another vector space over P, and let $\phi: G \to \mathrm{GL}(W)$ be a continuous homomorphism. Then, for every closed subset M of W, the group $\mathrm{Stab}_G(M) = \{g \in G | \phi(g)M = M\}$ is closed in G. In particular, the stabilizer of a point w in W is closed.

This list can easily be extended. Assertion (e) provides the best means of making concrete examples, except that we still lack examples of continuous homomorphisms. The following are the most typical of these.

(f) Let W be a subspace of V such that $GW = W$. Then, the restriction homomorphisms $G \to G|_W$ and $G \to G|_{V/W}$ are continuous.

(g) The action of a group G on $M_n(P)$ by left or right multiplication, and also by similarity transformations $x \to gxg^{-1}$ $(x \in M_n(P), g \in G)$, give continuous homomorphisms $G \to \mathrm{GL}(M_n(P))$.

(h) The natural action of a group G on the homogeneous components of the tensor, symmetric and exterior algebras of a space V (see 2.13, 2.14, 2.15) give continuous homomorphisms of G onto its images.

(i) If $\phi: G \to \mathrm{GL}_m(P)$ and $\psi: G \to \mathrm{GL}_r(P)$ are continuous homomorphisms, so is the tensor product $(\phi \otimes \psi): G \to \mathrm{GL}_{mr}(P)$.

We observe next that certain natural subgroups may not be Zariski closed. For example, $\mathrm{GL}_1(\mathbb{R})$ is not closed in $\mathrm{GL}_1(\mathbb{C})$ (because all closed subsets of the latter are finite). Similarly, $\mathrm{GL}_n(\mathbb{R})$ is not closed in $\mathrm{GL}_n(\mathbb{C})$ for any n. True, we can save the situation by representing complex numbers $a + bi$ by matrices $\begin{pmatrix} a & b \\ -b & a \end{pmatrix}$ over $\mathbb{R}$. This very natural map is not a continuous map from $\mathbb{C}$ to $M_2(\mathbb{R})$. Nevertheless, it enables us to equip $\mathrm{GL}_n(\mathbb{C})$ with a finer topology, namely that induced by the Zariski topology of $M_{2n}(\mathbb{R})$; and $\mathrm{GL}_n(\mathbb{R})$ is closed in this topology. A similar device can be very useful in a number of situations.

The following result of Chevalley has numerous applications.

Theorem. *Let H be a closed normal subgroup of a group G. Then, for some m, there exists a continuous homomorphism $\phi: G \to \mathrm{GL}_m(P)$ with kernel H.*

5.5. Connected Components. The concept of connectedness introduced in 5.2 is exceptionally useful in the study of linear groups in regard of the Zariski topology. It turns out that every group G has a unique connected normal subgroup of finite index. It is called the *connected component* (or component of the identity) of G and is denoted by G^0. If $G = G^0$, G is said to be *connected*. The *dimension* of a linear group is the dimension of its connected component G^0 in the sense of 5.2.

Thus, in some sense, the study of an arbitrary linear group G reduces to that of its connected component G^0 and the finite group G/G^0. Just as induction on the group order is so useful in studying finite groups, induction on the dimension is very effective for investigating connected groups.

Example. The group $D_n(P)$ of diagonal matrices is connected. The group $G = \mathrm{Mon}(P)$ of monomial matrices has $D_n(P)$ as a connected normal subgroup of index $n!$. Since G^0 is unique, we have $G^0 = D_n(P)$.

Further details on the elements of the Zariski topology can be found in the books of Kaplansky (1957), Dixon (1971), Wehrfritz (1973), and also in the introductory chapters of the monographs Borel (1969) and Humphreys (1975) devoted to algebraic groups.

§ 6. Algebraic Linear Groups

The construction of the theory of algebraic linear groups in the fifties and early sixties caused radical changes in the representation theory of linear groups. The theory of algebraic groups can be viewed as a natural generalization of the theory of classical groups (§ 3); at the same time, this theory is on an incomparably higher level, synthesizing as it does methods from group theory, number theory and algebraic geometry. The theory of algebraic linear groups is a self-standing branch of algebra. For this reason, we shall touch on it here only insofar as we need to for interpreting its interrelations with the general theory of linear groups. Nevertheless, we are forced to pay it a certain amount of attention.

In the modern context, the concept of algebraic group is fairly complex, and one can approach it on different levels. The most elementary of these is to set $G = \mathrm{GL}_n(P)$ as in 5.4, and (for an infinite field P), define an algebraic group $H \subset \mathrm{GL}_n(P)$ to be a subgroup closed in the topology explained in 5.4. If P is algebraically closed, the theory can be advanced fairly far and the principal structure theorems enunciated. We shall stick with this set-up, and return later to an explanation of the modern approach to the concept of algebraic group.

The theory of algebraic groups splits naturally into three more-or-less independent thrusts: structure theory, arithmetic theory and representation theory. The structure theory studies properties of the algebraic groups in the large and does not discriminate between isomorphic groups. In the main, the arithmetic theory studies arithmetic subgroups of algebraic groups over algebraic number fields (see 6.11); simplifying greatly, one can say that an arithmetic subgroup is the group of integer matrices in an algebraic matrix group (for instance, the groups $\mathrm{SL}_n(\mathbb{Z})$ and $\mathrm{Sp}_{2n}(\mathbb{Z})$). Finally, representation theory studies realisations of algebraic groups by matrices, that is, their representations. Each of these parts contains divisions that are of interest from the point of view of the general theory of linear groups. In 6.1–6.9 we shall discuss some results from the structure theory, the representation theory in 6.10, and in 6.11 we touch sketchily on some questions in the arithmetic theory.

6.1. Main Definitions. Throughout § 6, P is an algebraically closed field and $p = \mathrm{char}\, P$. A subgroup H of $\mathrm{GL}_n(P)$ is said to be *algebraic* if $H = M \cap \mathrm{GL}_n(P)$, where M is some Zariski-closed subset of $M_n(P)$ (see 5.4).

An algebraic group is said to be *connected* if it contains no closed proper subgroup of finite index; *reductive* if it has no unipotent normal subgroups; *semisimple* if it is connected and has no infinite abelian normal subgroups; *simple* if it is connected and contains no proper infinite closed normal subgroups; in particular, the centre of a simple algebraic group is finite, though it need not be trivial.

We explain at this point that a *unipotent* group is one consisting of matrices whose eigenvalues are 1; that is, matrices x such that $(x - E_n)^n = 0$. We shall see later that every unipotent group is similar to a subgroup of the upper unitriangular group (see 2.2.1).

The maximal connected soluble normal subgroup $R(G)$ of an algebraic group G is called its *radical*, and the maximal connected unipotent subgroup $U(G)$ its *unipotent radical*. A *Borel subgroup* is a maximal connected soluble subgroup; a *Cartan subgroup* is a maximal connected nilpotent subgroup having finite index in its normaliser. A connected abelian subgroup consisting of semisimple elements is called a *torus*. The *rank* of an algebraic group is the dimension of a maximal torus (in the sense of 5.2). In a semisimple group, every Cartan subgroup is a maximal torus, and conversely.

The theory of algebraic groups makes decisive use of the apparatus of algebraic geometry. Naturally, we cannot expound it here. However, it is absolutely imperative that we introduce the concept of morphism of algebraic groups. Let V and V' be two vector spaces over a field P, of dimensions m and n respectively. Choice of bases in them enables us to describe elements x in V and y in V' by their coordinates $(x_1, \ldots, x_n)$ and $(y_1, \ldots, y_m)$. A map $\phi\colon V \to V'$ is said to be *regular* if the coordinates of $\phi(x)$ are polynomials in $x_1, \ldots, x_n$, for every x in V. Let $M \subset V$ and $M' \subset V'$ be two Zariski closed sets; a map $\psi\colon M \to M'$ is said to be *regular* if ψ is the restriction to M of some regular map $\phi\colon V \to V'$. Finally, a *morphism* of algebraic groups is simultaneously a homomorphism $\phi\colon H \to H' \subset \mathrm{GL}_m(P)$ of abstract groups and a regular map of closed sets: $\{\mathrm{diag}(h, \det h^{-1}) | h \in H\} \subset M_{n+1}(P) \to \{\mathrm{diag}(\phi(h), \det \phi(h)^{-1}\} \subset M_{m+1}(P)$.

If a morphism is a one-to-one map whose inverse is also a morphism, we have an *isomorphism* of algebraic groups. A morphism $G \to \mathrm{GL}_n(P)$ of algebraic groups is called a (*rational*) *representation* of G.

6.2. The General Structure Theorem for Algebraic Groups

Theorem. *Let G be an algebraic group.* (1) *G has a connected normal subgroup of finite index.* (2) *If G is connected, then $G/R(G)$ is semisimple and $G/U(G)$ is reductive.* (3) *A reductive group is almost a direct product of a semisimple subgroup and the centre.* (4) *A semisimple group is almost a direct product of simple subgroups.* (5) *There are only finitely many simple groups with isomorphic central factor-groups. Among them there is a unique group, said to be simple-connected, such that all others are homomorphic images of it.*

In assertions (3) and (4), the word "almost" means that the pairwise intersections of the factors are finite. If $p = \mathrm{char}\, P = 0$, assertion (2) can be sharpened. Namely, a connected group is the semidirect product of a semisimple group and the radical, and also a semidirect product of a reductive group and the unipotent radical. We note also that, by Chevalley's theorem (see 5.4), $G/R(G)$ and $G/U(G)$ can be considered as algebraic groups in the sense of the definition adopted here.

6.3. Classification of Simple Algebraic Groups. Let G be a simple algebraic group, $G \subset \mathrm{GL}_n(P)$. Then $G = \phi(H)$, where ϕ is an algebraic group morphism with finite kernel contained in the centre of H, and H is one of the following groups: $\mathrm{SL}_m(P)$, $\mathrm{Sp}_{2m}(P)$, $\mathrm{Spin}_m(P)$ (see 3.12; in characteristic 2, m is even), $G_2(P)$, $F_4(P)$, $E_i(P)$ ($i = 6, 7, 8$). Groups of the last five types are called exceptional groups; the others are the classical groups. For these, the following notation is also accepted: $A_{m-1}(P)$ for $\mathrm{SL}_m(P)$, $C_m(P)$ for $\mathrm{Sp}_{2m}(P)$, $D_m(P)$ for $\mathrm{Spin}_{2m}(P)$, and $B_m(P)$ for $\mathrm{Spin}_{2m+1}(P)$. The lower index is the rank of the group H. Thus, from the point of view of the structure theory, there are four infinite families of simple algebraic groups and five exceptional types. Essentially, the description of realizations of them as subgroups of general linear groups is the central question of representation theory (see 6.9).

The reader familiar with the theory of Lie algebras will notice at once that the notation for simple algebraic groups is analogous to that for simple Lie algebras; more exactly, to the notation for their root systems. This is in no way accidental, of course. Every algebraic group is in canonical association with its Lie algebra. Moreover, every simple algebraic group has an associated root system; the root system of an algebraic group is identical with the root system of its Lie algebra. For arbitrary p, every simple algebraic group with trivial centre is the connected component of the automorphism group of its Lie algebra.

6.4. Conjugacy Theorems. Let H be an algebraic group, H a subgroup of $\mathrm{GL}_n(P)$. Groups of the following type are conjugate in H: (a) the maximal tori; (b) the Cartan subgroups; (c) the Borel subgroups; (d) the maximal connected unipotent subgroups (in characteristic 0, the connectedness condition is automatic); (e) the Sylow q-subgroups, for all primes q.

6.5. Structure of Algebraic Groups. The structure theory of algebraic groups is very fully developed, and we could easily extend the list of results in it. However, we shall mention only the following theorems, which have application in the theory of abstract linear groups. Let G be an algebraic group over an algebraically closed field P.

(A) Every element g of G can be expressed in the form $g_1 g_2$, where g_1 is *semisimple* (that is, diagonalisable), g_2 is unipotent and $g_1 g_2 = g_2 g_1$. If G is connected, every semisimple element lies in some maximal torus, and every unipotent element in some Borel subgroup.

(B) If G is reductive (or semisimple), the number of conjugacy classes of unipotent elements is finite.

(C) If G is semisimple and nontrivial, then G contains a subgroup isomorphic to $\mathrm{SL}_2(P)$ or $\mathrm{PSL}_2(P)$.

(D) Let G be a connected group, K a Cartan subgroup and N the normalizer of K in G. Then N/K is finite and is called the Weil group of G. The *Bruhat decomposition* holds: $G = \bigcup BwB$, where B is a Borel subgroup contaning K and w runs over a set of representatives of the cosets N/K.

(E) If G is connected, G is the union of the Borel subgroups. If G is reductive, the union of the maximal tori contains a nonempty open subset of G; the closure of the set of non-semisimple elements is different from G.

There is an exposition of the results of 6.1–6.5 in Humphreys (1975).

6.6. The Coordinate Ring of an Algebraic Group. With the notation of 5.1, let M be a set closed in the Zariski topology, and let I be the set of all polynomials in F that vanish on all the points of M. Then, the factor-ring $P[M] = F/I$ is called the *coordinate ring* of M. Furthermore, to every point $a = (a_1, \dots, a_k)$ in M there is a P-homomorphism $P[M] \to P$ induced by the map $f(x_1, \dots, x_n) \to f(a_1, \dots, a_n)$. The converse is also true: every such homomorphism defines a point of M. Thus, M is completely determined by its coordinate ring $P[M]$. In addition, M can be interpreted as the set $\mathrm{Hom}_P(P[M], P)$ of all P-homomorphisms from $P[M]$ to P.

It is natural at this point to pass to an algebraic point of view, that is, to investigate the structure of $\mathrm{Hom}_P(A, P)$ for an arbitrary commutative algebra A over F. Moreover, it is not necessary to assume here that P is a field; it can be replaced by any commutative ring with identity. This path leads us to the concept of scheme in algebraic geometry.

If $W = M_n(P)$ is the space of matrices and the subset M of W forms a group, then $P(M)$ has additional properties (such as a Hopf algebra structure). They are not difficult to write out explicitly: see § 7.6 of Humphreys (1975), for example. These properties can be taken as axioms; it turns out that the set $H = \mathrm{Hom}(P[M], P)$ is a group when they are satisfied. This gives a new definition of *algebraic linear group* that is nearer to the modern interpretation of the concept. Moreover, the conditions themselves make sense if P is any commutative ring with identity. In this way we arrive at the concept of *affine group scheme* over an arbitrary commutative ring. It is a curious fact that, for any commutative P-algebra B, the set $\mathrm{Hom}_P(A, B)$ of all P-homomorphisms from A to B forms a group. It is called the group of points of the scheme H over B. In this sense, the group scheme H can be viewed as a functor on the category of commutative P-algebras B with values in the category of groups. See § 1 of Borel (1966) for more details.

Let K be a subfield of the field P. We say that the ring $P[M]$ has a *K-structure* if $P[M]$ contains a K-subalgebra S that spans $P[M]$ over P, but no proper subalgebra S_1 of S has this property. For instance, the ring $K[x_{11}, \dots, x_{nn}]$ gives a K-structure on the coordinate ring $P[x_{11}, \dots, x_{nn}]$ of $\mathrm{GL}_n(P)$. As a rule there are many K-structures on $P[M]$.

6.7. Groups of K-points. We shall say that an algebraic group $G = \mathrm{Hom}_P(P(G), P)$ is *defined over* K if $P[G]$ has a (fixed) K-structure S. A group G with a fixed K-structure is called a *K-form*. In this situation, the set $\mathrm{Hom}_K(S, K)$

is a group; it is denoted by G_K and called the *group of K-points* (or sometimes the K-form) of G.

To avoid technical complexities, we shall assume that the field K is perfect. Fields of zero characteristic are all perfect. If $p = \text{char } K > 0$, then K is *perfect* if and only if $K = K^p$, where K^p is the subfield consisting of the p-th powers of elements of K.

Every ideal I of the K-algebra S defines a *K-closed subset* $M_I = \{g \in \text{Hom}_P(P[G], P) | g(I) = 0\}$ of G. In particular, a subgroup H of G is said to be a *K-subgroup* if $H = M_I$ for some ideal I of S. When we speak of K-subgroups (K-subsets) we shall always have in mind that the K-structure S of $P[G]$ is fixed (possibly, we should have chosen to write S_K-subgroup). Because of this agreement, we have $H_K \subset G_K$. The system of K-closed sets in G is called the *K-structure* of G.

If K is algebraically closed, there is essentially just one K-structure on G. Let K be an arbitrary field and P and algebraic closure of K. The classification problem for K-forms in this situation can be interpreted in terms of Galois cohomology: see Serre (1964), Chapter III, § 1. This approach leads in principle to classification of K-forms of simple algebraic groups (Tits, 1965). There are explicit classifications for a number of interesting fields. For instance, complete results have been obtained for finite and non-discrete locally compact fields.

We recall at this point an important theorem of Rosenlicht (1957) stating that, for an infinite (perfect) field K, G_K is dense in G in the Zariski topology.

We consider now a typical example of how to describe the group of K-points by introducing different K-structures into the coordinate ring $P[G]$. Suppose that $P = \mathbb{C}$, $K = \mathbb{R}$, and $G = \{\text{diag}(a, a^{-1})\} \subset \text{GL}_2(\mathbb{C})$, a running over the nonzero elements of $\mathbb{C}$. The coordinate ring $\mathbb{C}[G]$ of the group G, which is isomorphic with $\text{GL}_1(\mathbb{C})$, is $\mathbb{C}[x_{11}, x_{22}]/(x_{11}x_{22} - 1) \cong \mathbb{C}[t, t^{-1}]$, where $(x_{11}x_{22} - 1)$ is the ideal of $\mathbb{C}[x_{11}, x_{22}]$ generated by $x_{11}x_{22} - 1$. The ring $\mathbb{C}[t, t^{-1}]$ has the obvious $\mathbb{R}$-structure $\mathbb{R}[t, t^{-1}]$ corresponding to the group isomorphic to $\text{GL}_1(\mathbb{R})$. Next, we consider the subring S of $\mathbb{C}[t, t^{-1}]$ consisting of the functions fixed by the automorphism $f(t) \to \overline{f(t^{-1})}$ (the bar means taking the complex conjugate of each coefficient of f). It is clear that $f(a) \in \mathbb{R}$ if $f \in S$ and $a = 1$ ($a \in \mathbb{C}$). In fact, there are no other points a in $\mathbb{C}$ such that $f(a) \in \mathbb{R}$ for all $f \in S$. In other words, $\text{Hom}_{\mathbb{R}}(S, \mathbb{R})$ is interpreted as the group $\mathbb{C}_1$ of complex numbers of modulus 1. Thus, $\mathbb{C}_1$ is the group of $\mathbb{R}$-points (the $\mathbb{R}$-form) of $\text{GL}_1(\mathbb{C})$. We note that $C_1 = U_1(\mathbb{C})$, the group of unitary matrices of degree 1 (see 3.7). In the same way, it can be shown that $U_n(\mathbb{C})$ is the $\mathbb{R}$-form for $\text{GL}_n(\mathbb{C})$. Moreover, for an arbitrary field P with an automorphism σ of order 2 (so that $\sigma^2 = 1$) and every Hermitian $n \times n$ matrix Γ, $U(\Gamma, P)$ is the group of K-points of $\text{GL}_n(P)$, where K is the fixed-point field of σ in P.

We mention here the following two important types of K-forms of simple algebraic groups. A group G is said to be *K-split* if some maximal torus has the direct product structure $H_1 \times \cdots \times H_m$, where the H_i are K-subgroups isomorphic to $\text{GL}_1(P)$. If G is K-split, then G_K has structure similar to that of G (G_K can be given by a set of generators and relations like those of G: see

[Steinberg (1967) § 6]. A group G is said to be *quasi-split* over K if one of its Borel subgroups is a K-group. Every split group is quasi-split. If K is finite, every group G defined over K is quasi-split. The unitary groups $U(f, V)$ are not split; they are quasi-split if and only if the index of f is $[\dim V/2]$ (integer part).

Let R be a subring of K with field of fractions K. The group G_R is defined like G_K, that is, by means of an R-structure on $P[G]$. Even if G_K is fixed, G_R is not uniquely determined. If G_K is given in terms of matrices over K, we can use a simpler verson by setting $G_R = G_K \cap \mathrm{GL}_n(R)$.

6.8. Structure of Groups of K-points of Simple Groups. Most of the important linear groups turn out to be groups of K-points of algebraic linear groups. Moreover, many of the structure problems for these groups have a common interpretation and a unifom solution in terms of algebraic groups. There is less information when K is a general ring. However, for an arbitrary field K, the structure theory is fairly well developed. We shall restrict ourselves here to some qualitative indications which may enable the reader to orient himself in this important area of algebraic group theory. A more rounded impression can be obtained through acquaintance with Borel (1969); see also Borel and Tits (1965).

Let G be a simple algebraic linear group defined over a field K. If G_K contains non-identity unipotent elements, we say that G_K is *isotropic*; otherwise G_K is *anisotropic*. The problem of describing the normal subgroups of anisotropic groups G_K that are not closed in the Zariski topology is exceedingly difficult, and indeed this subject is being intensively studied at present. It is completely resolved in the case where K is a non-discrete locally compact field (perhaps we should recall that G_K is compact in this case). See [Zalesskij (1981) § 8] for more details.

The following theorem of Tits (1964) is the most general result on the structure of isotropic groups.

Let H be the subgroup of the isotropic group G_K generated by all the unipotent elements of G_K, and Z its centre. Then H/Z is abstractly simple except where $G_K = \mathrm{SL}_2(2), \mathrm{SL}_2(3), {}^2G_2(3), {}^2F_4(2)$.

This theorem incorporates many results on the structure of classical groups (Dieudonné (1971)).

What can be said about G_K/H? This problem splits into two problems of different natures. The first, called the Kneser-Tits problem, asks for a description of G_K/H when G is simply-connected. The second is to describe the groups $G_K/\phi(\tilde{G}_K)$, where $\tilde{G}$ is the simply-connected cover of G and ϕ the canonical homomorphism $\tilde{G} \to G$ (the *isogeny*).

The conjecture that $G_K = H$ for simply-connected groups G (posed for $G = \mathrm{SL}_n$ by Tannaka and Artin in the forties) has been proved false (Platonov, 1975). Note that for simply-connected G, G_K/H is always an abelian group of finite exponent. It turns out that for $G = \mathrm{SL}_n$, $n > 1$, and for a suitable K there exists a K-form G_K such that G_K/H is practically any abelian group of finite exponent. For the field of algebraic numbers, as well as for finite fields, $G_K = H$ when G is simply-connected.

Suppose that G is not simply-connected, that is, $G \neq \tilde{G}$. For finite K, we have the precise formula $|\mathrm{Ker}(\tilde{G} \to G)| = |G_K/\phi(\tilde{G}_K)|$. The most general result is that $\phi(\tilde{G}_K)$ is a proper subgroup of G_K for any finitely generated field K (Platonov 1974).

6.9. Chevalley Groups. The classical definition of the Chevalley groups uses the apparatus of Lie algebras. However, at the present stage of its development, the theory of the Chevalley groups can only be expounded adequately from the point of view of algebraic group theory. It has to be said that most combinatorial results can be deduced from some of the simplest properties of the Chevalley groups, the Bruhat decomposition being central here (see 6.5); this is where the theory of the Chevalley groups grows into the theory of groups with BN-pairs, which is a comparatively new but fairly well-advanced branch of group theory.

Let K be a subfield of the field P and G a quasi-split simple (or sometimes, we just assume semisimplicity) group over K. By a *Chevalley group over* K we understand a subgroup H of G_K generated by unipotent elements. If G is simply-connected, then $H = G_K$, and in this case H is called a *universal Chevalley group*. By Tits's Theorem (see 6.8), every proper normal subgroup of a Chevalley group is central (with the exception of the cases enumerated above).

The Chevalley groups are divided into the normal or twisted types according as to whether G_K is split or not. For example, $\mathrm{SL}_n(K)$ is of normal type and $\mathrm{SU}_n(K)$ is twisted.

If K is finite, we get *finite Chevalley groups*. At first sight it seems unlikely that the theory of algebraic groups, which imitates the theory of Lie groups to some extent, could be useful in the study of finite groups. It was discovered by Chevalley in 1955 and made more evident in works of his successors that algebraic groups are indeed powerful tools for investigating finite groups. Moreover, finite Chevalley groups (sometimes called finite groups of Lie type) comprise the overwhelming majority of finite nonabelian simple groups. More exactly, the only other finite nonabelian simple groups are the alternating groups and the 26 so-called *sporadic groups.*

Every simple algebraic group G is defined over the prime subfield P_0 of P. Because of this, Chevalley groups of normal type exist for every subfield K of P. This is not the case for the twisted groups. The finite twisted Chevalley groups are exhausted (up to the centres) by the following groups: ${}^2A_n(q)$ (these group are isomorphic to the central quotients of $\mathrm{SU}_{n+1}(q^2)$); ${}^2D_n(q)$, which is the commutator subgroup of the orthogonal group of even dimension $2m > 6$ corresponding to a quadratic form of index $m - 1$; and the five exceptional series ${}^2E_6(q)$, ${}^3D_4(q)$, ${}^2G_2(q)$, ${}^2F_4(q)$, ${}^2C_2(q)$, the last two of which exist only in characteristic 2, while ${}^2G_2(q)$ exists only in characteristic 3. Here $q = |K|$.

Many questions about the structure of finite Chevalley groups have been solved very effectively, thanks to the convenient relations between the so-called root subgroups (Steinberg (1975), § 6).

The realisation of the Chevalley groups by matrices over a field P is in essence that part of the representation theory of the Chevalley groups which is developed within the confines of the representation theory of algebraic groups over P. The key fact is that every irreducible representation of a Chevalley group H in $GL_n(P)$ extends to a representation of the corresponding algebraic group G (Steinberg (1967)). In contrast, the representation theory of finite Chevalley groups by complex matrices cannot be based on the general representation theory of algebraic groups; a different technique is developed here, in which many first-state ideas were borrowed from the representation theory of Lie groups by unitary operators in Hilbert space. The prime aim, namely the construction of the irreducible characters of the finite Chevalley groups, is in all probability close to completion. Explicit tables of characters exist for groups of rank at most two. For the general case, the idea is to find explicit formulae and to develop machinery that can find them according to a certain scheme.

Comparatively little is known about the representations of Chevalley groups over fields of characteristic $q \neq 0, p$, where $p = \text{char } K$.

The main sourcebooks on the theory of the Chevalley groups are those of Carter (1972) and Steinberg (1967). The representation theory of the Chevalley groups over P is expounded in Steinberg (1967), and partially in the Seminar on Algebraic Groups and Related Subgroups (1970). Steinberg (1967) is also a place where the representation theory over $\mathbb{C}$ can be found; the current stage of development of representation theory over $\mathbb{C}$ is expounded in Carter (1985) and Lusztig (1984).

6.10. Representation Theory of Algebraic Groups. The most important part of the representation theory of algebraic groups is that relating to simple groups. Practically no important regularities have been established for non-reductive groups. If P is of characteristic 0, every representation of a reductive group is completely reducible. Moreover, a representation of an arbitrary group G is completely reducible if and only if the unipotent radical $U(G)$ (see 6.1) is contained in the kernel of the representation. In particular, every irreducible representation of a group G can be realized *via* a representation of a reductive group, namely $G/U(G)$.

If $\text{char } P = p > 0$, every representation of a reductive group is completely reducible if and only if the connected component of G is abelian and the factor-group by it contain no element of order p. However, for $p > 0$ there exists a weak form of complete reductivity called "geometric reductivity" (this means that when G acts on a polynomial ring $S = P[x_1, \ldots, x_n]$ (see 2.14) in such a way that if the space $\langle x_1 \rangle$ is invariant under G, then for suitable r the subspace $\langle x_1^r \rangle$ has a G-invariant complement in the space $S_{(r)}$ of homogeneous polynomials of degree r). This fact is the basis of the proof of the Hilbert-Mumford Theorem in invariant theory stating that $S^G = \{s \in S \mid gs = s \text{ for all } g \text{ in } G\}$ is finitely generated as an algebra over P.

If G is connected and reductive, then G is almost a direct product (see 6.2) of a semisimple group and a central connected abelian subgroup. This description of the irreducible representations of G comes down to a description of the representations of the semisimple part.

Basically, the representation theory of semisimple groups studies irreducible representations. It is certainly true in characteristic 0 that everything reduces to that case (because of complete reducibility), whereas this is far from the truth for $p > 0$. For $p > 0$, almost everything is known about representations of $SL_2(P)$. For other groups, results are significantly less complete.

The description of the irreducible representations of semisimple groups reduces to the simple group case, because of assertion (4) in 6.2 together with elementary facts from representation theory. We shall assume in what follows that G is simple, one of the groups listed in 6.3.

Thus, let G be a simple simply-connected aglebraic group of rank r, say. The irreducible (rational) representations of G are parametrised by "highest weights", which can be interpreted as vectors $\omega = \alpha_1\omega_1 + \cdots + \alpha_r\omega_r$ with non-negative integer coordinates $\alpha_1, \ldots, \alpha_r$. The representations ϕ_i corresponding to the basis vectors ω_i ($i = 1, \ldots, r$), are the *fundamental* representations. The unique one-dimensional representation corresponds to the zero vector. Every irreducible representation is constructed in a definite way from the fundamental representations. More exactly: a representation ϕ with highest weight ω is a quotient of the tensor product $\otimes^{\alpha_1}\phi_1 \otimes^{\alpha_2} \cdots \otimes^{\alpha_r}\phi_r$ (each ϕ_i occurs α_i times). If $p > 0$, then P has the *Frobenius automorphism* $\sigma\colon \alpha \to \alpha^p$ ($\alpha \in P$). We denote the extension of σ to $GL_n(P)$ by the same letter: under it, every matrix entry is taken to its p-th power. If the Frobenius automorphism is applied to $\phi(G)$, we get a representation of highest weight $p\omega$. Therefore, the degrees of all the representations with highest weights $p^l\omega$ ($l = 1, 2, \ldots$) are equal. Let $\mathfrak{R}_p$ be the set of all irreducible representations with bounded highest weights, that is, with $\alpha_i < p$ for $i = 1, \ldots, r$. Then, the tensor product $\sigma^{t_1}\lambda_1 \otimes \sigma^{t_2}\lambda_2 \otimes \cdots \otimes \sigma^{t_m}\lambda_m$ is irreducible for all choices of pairwise different non-negative integers $t_1, \ldots, t_m$ and all $\lambda_1, \ldots, \lambda_m$ in $\mathfrak{R}_p$ (not necessarily different this time). Moreover, every irreducible representation can be constructed in this way from those in $\mathfrak{R}_p$.

It must be said that the representation theory of simple algebraic groups over a field of characteristic 0 is completely parallel to the representation theory of simple Lie algebras, and also to the theory of finite-dimensional representations of simple Lie groups. Therefore, many proofs of properties of the representations of algebraic groups can be deduced from sources in the representation theory of Lie algebras and Lie groups (see Zhelobenko and Shtern (1983), for example).

It is a remarkable fact that every irreducible representation of a simple algebraic group can be constructed from the rational irreducibles. More exactly, let G be a simply-connected algebraic group and K an infinite field. Let ϕ be any non-trivial irreducible representation of G over an algebraically closed field P. Then there exists a finite set of embeddings $\rho_i\colon K \to P$ and a set of rational representations ϕ_i of G_K such that $\phi = \bigotimes_i \rho_i\phi_i$ (Borel and Tits (1973); see also Theorem 42 of Steinberg (1975). It follows from this that K can be embedded in P; in particular, the two fields have the same characteristic. It is clear that this theorem does not carry over to the case where K is finite. However, if char K = char P, the result still holds (see Steinberg (1975)).

The existence of a classification of the irreducible representations in no way means that we have effective methods for solving every question about the irreducible representations of simple groups. The availability of comparatively well-developed machinery notwithstanding, even in characteristic 0 the study of properties of concrete representations normally requires highly laborious work. A broad sweep of information about the representations with given highest weight can be extracted from the weight table. The *weights* of a representation ϕ are the elements of $\mathrm{Hom}(T, P^*)$ arising when ϕ is restricted to a maximal torus T of G. We have $\mathbb{Z}^r \cong \mathrm{Hom}(T, P^*)$, where r is the rank of G. The weights can therefore be written as integral vectors. It turns out that the set of weights of a representation ϕ does not depend on the choice of maximal torus T. The weights $\omega_1, \ldots, \omega_n$ given by $\omega_i = (\underbrace{0, \ldots, 0}_{i-1}, 1, 0, \ldots, 0)$ are the *fundamental* weights. For a connected simple algebraic group over a field of characteristic 0, the weight table of an irreducible representation ϕ with highest weight $\omega \in \mathbb{Z}^r$ is exactly the same as the weight table of the simple Lie algebra of the same type as G and having highest weight ω. For algebraic groups over fields of characteristic $p > 0$, this is no longer true. However, for representations with bounded highest weights (the so-called infinitesimal irreducibles), the above assertion holds except in the cases $p = 2$, G is of type $B_r \cong C_r$ or F_4, and where $p = 3$ and G is of type G_2 (Premet, 1987).

There is a canonical action of an algebraic group G on its Lie algebra (see 6.3); it is called the adjoint representation of G. The nonzero weights of the adjoint representation are called the roots. With every basis $\omega_1, \ldots, \omega_r$ of a system of weights there is associated a set of roots $\alpha_1, \ldots, \alpha_r$ called the simple roots. It turns out that every root is an integer linear combination of simple roots, the coefficients being either all positive or all negative. Every weight of an irreducible representation ϕ is of the form $\omega - \sum m_i\alpha_i$, where the m_i are non-negative and ω is a highest weight (indeed, this is the explanation of the term "highest weight"). From the point of view of linear group theory, the most important invariant of a representation is dimension. In the characteristic zero case, there are formulae expressing dimensions in terms of highest weights. The construction of such formulae for fields of positive characteristic is one of the fundamental problems in the representation theory of algebraic groups. The main sourcebooks on the representation theory of algebraic groups of prime characteristic are those of Steinberg (1975) and Humphreys (1976).

6.11. Arithmetic Theory of Algebraic Groups and Linear Groups over Arithmetic Rings. Roughly speaking, the objects of study in the arithmetic theory of algebraic groups are groups defined over global fields (that is, fields of algebraic numbers; in particular, $\mathbb{Q}$ and fields of algebraic functions of a single variable with finite or locally finite field of constants). Let L be a global field, Z its ring of integes, L_v the completion of L in some valuation v, and Z_v the valuation ring of L if v is non-Archimedean (see 4.6). Let G be an algebraic linear group defined over L. The arithmetic theory investigates the connections between the groups

G_L, G_Z, G_{L_v}, G_{Z_v}, and of course the groups themselves. The ideas of algebraic number theory have had a profound influence on the arithmetic theory. We shall restrict attention to a discussion of questions closely connected with the general theory of linear groups. The reader will find the fullest available exposition of the main ideas of the arithmetic theory in the book of Platonov and Rapinchuk (1991). Humphreys's book (1980) is an elementary introduction to the theory.

The following concept is pivotal for the entire theory: a subgroup A of G_L is said to be *arithmetic* if it is *commensurable* with G_Z, that is, if $A \cap G_Z$ is of finite index in A and in G_Z. The group G_Z is not uniquely determined (see 6.6), but the arithmeticity property of A does not depend on the choice of G_Z in G_L (G_L is fixed here). In particular, all the G_Z are mutually commensurable.

Theorem (Borel and Harish-Chandra 1962). *If L is a global field of characteristic* 0, *every arithmetic subgroup A of G is finitely presented.*

This theorem is valid for almost all global fields of characteristic $p > 0$; however, if k is a finite field, $\mathrm{SL}_2(k[x])$ is not finitely generated.

The classical problem of describing the normal subgroups of $\mathrm{SL}_n(\mathbb{Z})$ has seen very significant development within the arithmetic theory of algebraic groups. For $n = 2$, these normal subgroups are hard to discern, since $\mathrm{SL}_2(\mathbb{Z})$ contains a free subgroup of finite index. On the other hand, for $n > 2$ every normal subgroup is central modulo some congruence subgroup. We recall that the congruence subgroup of level I, where I is an ideal of $\mathbb{Z}$, is that consisting of the matrices g in $\mathrm{SL}_n(\mathbb{Z})$ that are congruent to the identity modulo I. The general problem is to describe the normal subgroups of all arithmetic subgroups of algebraic groups defined over global fields. The reader will find surveys of the investigations relating to this theme in papers of Platonov (1982) and Platonov and Rapinchuk (1983), for example. We shall provide here a few results so that the reader can estimate for himself the level of generality of the most recent achievements. We shall need the concept of L-rank of a group G, which generalizes the concept of index in classical groups. If G is an algebraic group defined over L, $G \subset \mathrm{GL}_n(L)$, its *$L$-rank* is the dimension of its "diagonal L-part" $D_n(L) \cap G$; more exactly, it is the largest number r such that G_L contains a subgroup isomorphic to the direct product of r copies of $\mathrm{GL}_1(L)$. It turns out that if the sum of the L-ranks of a simple algebraic group G over all Archimedean valuations v of L is not less than 2, then every noncentral normal subgroup of G_Z is of finite index in it (Margulis 1978). This is fairly close to the natural limit provided by $\mathrm{SL}_2(\mathbb{Z})$: here $G = \mathrm{SL}_2(\mathbb{Q})$, $\mathbb{Q}$ has just one Archimedean valuation up to equivalence (namely, the absolute value), so that the sum just mentioned is 1. The field $\mathbb{Q}(\sqrt{2})$ has two inequivalent Archimedean valuations (each obtained from the other under the Galois automorphism $a + b\sqrt{2} \to a - b\sqrt{2}$ $(a, b \in \mathbb{Q})$; thus, Margulis's Theorem applies to $\mathrm{SL}_2(\mathbb{Z}\sqrt{2})$. We note here that there are noncentral normal subgroups of infinite index in $\mathrm{SL}_2(\mathbb{Z}\sqrt{d})$, for every negative integer d.

The problem of describing the (normal) subgroups of finite index in arithmetic groups is known as the *congruence problem*. Its simplest form is this: is it true that every subgroup of finite index contains a *congruence subgroup*? These subgroups are defined just as for $SL_n(\mathbb{Z})$ above: if I is an ideal of Z, the *congruence subgroup of level I* consists of the matrices g in G_Z congruent to the identity modulo I. Therefore, the congruence subgroup is not uniquely defined, and like G_Z, it depends on the realization of G_L by matrices over L. However, this does not influence the solution of the congruence problem, since any two congruence subgroups are commensurable. In contemporary language, the congruence problem is understood to mean the computation of the so-called congruence kernel. This is how it is defined: for an arithmetic group A, let $\hat{A}$ be its completion in the topology defined by all the subgroups of finite index, and $\bar{A}$ the completion in that defined by all congruence subgroups (both topologies are separable since the congruence subgroups intersect in the identity). The kernel $C(A)$ of the canonic homomorphism $\hat{A} \to \bar{A}$ is called the *congruence kernel*. It is a profinite group characterising the set of subgroups of A of finite index that are not congruence subgroups. It is usual to distinguish three cases: $|C(A)| = 1$, in which case every subgroup of finite index contains a congruence subgroup; $|C(A)|$ finite; $|C(A)|$ infinite. The congruence kernel has been described in recent years for many types of arithmetic groups (see Platonov and Rapinchuk (1983)). The following classical result is due to Bass, Milnor and Serre (1967).

If $A = SL_n(\mathbb{Z})$ or $Sp_n(\mathbb{Z})$, $n > 2$, then $C(A) = 1$ except in the case where L cannot be embedded in $\mathbb{R}$, when $C(A) = \mu_L$, the group of roots of unity in L.

For $n = 2$, the determination of the congruence kernel is a very difficult problem. It was solved for $SL_2(\mathbb{Z})$ by Mel'nikov (1976): he proved that, in this case, the congruence kernel is a free profinite group of countably infinite rank. There are some results for the quadratic extensions $L = \mathbb{Q}(\sqrt{-m})$, $m \in \mathbb{N}$.

Chapter 2
The Machinery of Linear Group Theory

§1. Analysis of Linear Groups by Elementary Linear Algebraic Methods

1.1. A Preliminary Analysis of Linear Groups. Practical problems do not often lead to linear groups with properties and structure that are self-evident. To get an idea of what a group arising for some reason or another actually is, much hard work is frequently necessary even if the end-result is a simple answer. With what should we begin, and what problems is it desirable to solve in the first place? The overall approach in algebra is to represent the object of study as a composite of simpler parts. In the case of linear groups, it would be pleasant, for instance, to discover that the group we are dealing with is similar to a diagonal or a triangular group, *etc.* In order to give a natural formulation of problems in a preliminary analysis of the situation, it is convenient to use "geometric" terminology, that is, terminology arising from the action of the group on a vector space. The development of the techniques arising from this approach has led eventually to the theory of modules. We shall not need to use any deep-lying facts from this theory; however, the language is more convenient than that of matrix theory. Although the general theory considers only modules over rings, for simplicity of expression we shall take the liberty of speaking about L-modules in cases where L is any set of linear maps. A typical situation we shall meet is the following. Let V be a vector space of finite dimension over a field P, and L a set of linear maps. We shall be interested in subspaces W of V such that $LW = W$. Every such subspace will be called an *L-module*. The orthodox terminology here is $[L]_P$-module, where $[L]_P$ is the *enveloping algbra* of L, that is, the smallest P-subalgebra of $\mathrm{End}_P V$ containing L. It is clear that every L-module is an $[L]_P$-module, and conversely. If W_1 and W_2 are L-modules such that $W_1 \subset W_2 \subset V$, the elements of L induce linear maps in the quotient-space W_2/W_1. With this in mind, we shall speak of the L-module W_2/W_1 and use the term "quotient-module" in this context.

1.2. Decomposable Modules. Fix some subset L of $\mathrm{End}_P V$. We shall say that V is a *decomposable* L-module if there is a decomposition $V = W_1 + W_2$ in which W_1 and W_2 are nonzero L-modules and $W_1 \cap W_2 = \{0\}$. Otherwise, V is said to be *indecomposable*. Because of the natural correspondence between matrices and linear maps, an L-module V is decomposable if and only if L can be given by block-diagonal matrices with respect to a suitable basis, with fixed set of blocks, of course. For example, if L is a set of matrices of the form $\begin{pmatrix} a & b \\ b & a \end{pmatrix}$, where $a, b \in \mathbb{C}$, then V is decomposable as L-module since L is similar to a set of diagonal matrices of the form $\begin{pmatrix} * & 0 \\ 0 & * \end{pmatrix}$. As a rule, detection of the decomposability of an L-module reduces the investigation of the properties of the action of L on V to those of the action of L on W_1 and W_2. It should be observed that decomposability is equivalent to the existence of an element $e \in \mathrm{End}_P V$ such that $e^2 = e$, $e \neq 0, 1$, and $el = le$ for all l in L.

A decomposition $V = W_1 + W_2$ can be refined if W_1 or W_2 is a decomposable L-module. A decomposition $V = W_1 + \cdots + W_n$ as a direct sum of L-modules is said to be complete if all the W_i are indecomposable. Every complete decomposition can be taken to any other by the action of some element g of $\mathrm{GL}(V)$ that commutes with every element of L (this is the Krull-Schmidt Theorem).

1.3. Reducible Modules. A more important concept in linear group theory is that of reducibility. A set L and an L-module V are said to be *reducible* if there is a nontrivial (that is, nonzero and proper) L-submodule W of V, and *irreducible* otherwise. It is easy to see that V can be simultaneously reducible and indecomposable. An L-module V and set L are said to be *completely reducible* if the components of a complete decomposition are irreducible L-modules.

If W is a reducible L-module, the series $\{0\} \subset W \subset V$ of L-modules can be refined: $\{0\} \subset W_1 \subset W \subset V$. This can also be done when V/W is reducible. It is easy to see that a composition series (that is, unrefinable series) of L-modules $\{0\} = W_0 \subset W_1 \subset \cdots \subset W_r = W$ can always be constructed. The factors W_i/W_{i-1} are irreducible L-modules for $i = 1, \ldots, r$, and they are called *composition factors*. Every two composition series have the same set of composition factors, up to the order of appearance and isomorphism as L-modules (this is the Jordan-Hölder Theorem).

If a basis consonant with a complete decomposition (with a composition series respectively) is chosen in V, then L takes the form

$$\begin{bmatrix} L_1 & 0 & \ldots & 0 \\ 0 & L_2 & \ldots & 0 \\ \multicolumn{4}{c}{\dotfill} \\ 0 & 0 & \ldots & L_r \end{bmatrix}, \qquad \begin{bmatrix} L_1 & * & \ldots & * \\ 0 & L_2 & \ldots & * \\ \multicolumn{4}{c}{\dotfill} \\ 0 & 0 & \ldots & L_r \end{bmatrix},$$

where $L_1, \ldots, L_r$ are indecomposable (irreducible respectively) sets of matrices. This is, of course, consistent with 1.2.2.

The reader must be warned: it does not follow that a set L of matrices is irreducible if it is not of block-diagonal form; what is required is that it cannot be brought to such a form by a similarity transformation. This remark is not superfluous, since there are important results about irreducible sets that are not true in general. The following celebrated result is a typical example.

Schur's Lemma. *Every nonzero matrix that permutes with all the matrices in some irreducible set L is invertible. An equivalent statement: the set L of matrices that permute with a given nonzero singular matrix is reducible.*

For, if M is a nonzero singular matrix, then $\{0\} \neq MV \neq V$. For $l \in L$ we have $lMV = MlV \subset MV$, that is, MV is a nontrivial L-module.

Schur's Lemma serves as a convenient test for the reducibility of a set (or group) of matrices, since finding a matrix M permuting with all the matrices in a set L comes to solving a system of homogeneous equations; this then gives the nontrivial module MV in explicit form. Unfortunately, the absence of nonzero singular matrices permuting with all members of L does not guarantee irreducibility of L; an example is the set $L = \left\{\begin{pmatrix} a & c \\ 0 & b \end{pmatrix}\right\}$, where $a, b \in \mathbb{C}$.

The property that L be irreducible depends on the ground field P. For example, $\begin{pmatrix} 0 & 1 \\ -1 & 0 \end{pmatrix}$ is irreducible over $\mathbb{R}$ and reducible over $\mathbb{C}$. A subset L of $M_n(P)$ is said to be *absolutely irreducible* if it is irreducible over every extension of P. If P is algebraically closed, every irreducible set is abolutely irreducible. There is a sharpened version of Schur's Lemma, namely: every matrix that permutes with each matrix in an absolutely irreducible set is scalar.

1.4. Imprimitive Modules. A set L and an L-module V is said to be *imprimitive* if there is a decomposition $V = W_1 + \cdots + W_m$ of V as a direct sum of proper subspaces W_i such that $lW_i = W_{j(l)}$ for all $l \in L$, $i \in \{1, \ldots, m\}$, and *primitive* otherwise. We shall employ the definition only in cases where L is an irreducible group. It is easy to check that an irreducible group G is imprimitive if and only if in some basis it is a group of block-monomial matrices (see 1.1.6). The following assertion is less trivial. An irreducible group G is imprimitive if and only if the identitification representation $G \to G$ is an induced representation (see 1.2.12). For the subgroup H as in 1.2.12 we must take $\{g \in G | gW_1 = W_1\}$, and for ϕ the restriction $H|_{W_1}$.

The set of spaces W_i is called an *imprimitivity system* for G.

§2. Machinery of the Theory of Algebras in Linear Group Theory

2.1. Burnside's Theorem and its Consequences. Many of the results in the theory of algebras that are applied in investigating the structure of linear groups are classical and were discovered more than 50 years ago. Such results are used largely to establish the reducibility or imprimitivity of linear groups. The corresponding techniques are pretty well complete by now. The theorems of Burnside and Clifford play a central role here.

Burnside's Theorem. *Let G be an absolutely irreducible subgroup* (*see* 1.3) *of* $\mathrm{GL}_n(P)$. *Then* $[G]_P = M_n(P)$; *in particular, G contains n^2 linearly independent matrices.*

The second assertion follows from the first since the enveloping algebra of G is the linear span $\{\sum \alpha_i g_i | \alpha_i \in P, g_i \in G\}$. We note that the first result holds for every absolutely irreducible set. The next lemma, also due to Burnside, has manifold application.

Lemma. *Let G be as in the theorem. The system of equations* $\mathrm{tr}(xg) = \alpha_g$, *where g runs over G and each α_g is an element of P, has at most one solution in G.*

This follows from Burnside's Theorem and the well-known fact that the bilinear form $\mathrm{tr}(xy)$ on $M_n(P)$ is nondegenerate.

Corollary A. *Let G be an absolutely irreducible group. If the set* $\mathrm{tr}(G)$ *of traces of elements of G consists of just one element, then $|G| = 1$ and $n = 1$. If* $\mathrm{tr}(G)$ *is finite, then G is finite. If it is countably infinite, so is G.*

We can go further. If P has a topology (metric or valuation) and this is extended in the natural way to $M_n(P)$, then $\overline{G}$ is compact (bounded, respectively) if $\overline{\mathrm{tr}}(G)$ is compact (if $\mathrm{tr}(G)$ is bounded, respectively).

Corollary B. *Every unipotent group of matrices is similar to a subgroup of the group of upper unitriangular matrices.*

We recall that a matrix is *unipotent* if all its eigenvalues are 1. A group consisting of unipotent matrices is also said to be *unipotent*. We embed the ground field in an algebraically closed field, and then our group reduces to block-triangular form with absolutely irreducible diagonal blocks (see 1.2).

By Corollary A, the blocks are one-dimensional. This proves the corollary for algebraically closed fields. In particular, there exists a vector fixed under all the elements of the group. It is obtained by solving a system of linear equations with coefficients in P. Since the system has a solution over the closure of P, it has a solution in P. Now simply proceed by induction.

A similar argument establishes the next result.

Corollary C. *Let G be a linear group such that $g^m = 1$ for all g in G. If* char $P = 0$, *then G is finite. If* char $P = p > 0$, *G is a finite extension of a p-group.*

2.2. Clifford's Theorem. The aim here is to find imprimitivity conditions for linear groups in terms of normal subgroups. We recall that an irreducible group $G \subset \mathrm{GL}(V)$ is said to be imprimitive if $V = V_1 + \cdots + V_k$, a direct sum of subspaces permuted amongst themselves by the action of G, with $k > 1$.

Theorem (Clifford, 1937). *Let G be an irreducible subgroup of* GL(V) *and N a proper normal subgroup. Then N is completely reducible. If $V_1, \ldots, V_k$ are the sums of mutually equivalent irreducible N-submodules, then G permutes the V_i among themselves transitively. In particular, if $k > 1$, G is imprimitive.*

Proof. Let V_0 be the sum of the irreducible N-submodules of V. Then $GV_0 = V_0$. Since G is irreducible, we have $V_0 = V$, and thus $V = V_1 + \cdots + V_k$. It follows (not quite trivially) that N is completely reducible. By the Jordan-Hölder Theorem (see 1.3), the V_i are permuted amongst themselves by the action of G.

This argument should be compared with the proof of the following assertion.

(*) Under the assumptions of Clifford's Theorem, the algebra $[G]_P$ is simple and $[N]_P$ is semisimple. If $[N]_P$ is not simple, then G is imprimitive.

Proof. Let R be the radical of $[G]_P$. Then $R^n = \{0\}$, where $n = \dim V$. Thus $RV \neq V$, and if $R \neq \{0\}$, RV is a nonzero G-module, contradicting the irreducibility of G. Thus, $R = \{0\}$. If $[G]_P$ is not simple, it is a direct sum of simple algebras. Let $E_n = e_1 + \cdots + e_k$ be a decomposition of the identity as the sum of the identities of the simple components. Then $e_i V$ is a nontrivial G-module, contrary to assumption. Further, if S is the radical of $[N]$, we have $gSg^{-1} = S$ for all g in G. It follows that $(SG)^n = \{0\}$ (since $S^n = \{0\}$), where SG is the ideal of $[G]$ generated by S. But then $SG \subset R = \{0\}$ and $S = \{0\}$, that is, N is semisimple. If it is not simple, the decomposition of its identity as the sum $E_n = e_1 + \cdots + e_r$ of the identities of its simple components leads to a decomposition of V into the imprimitivity system $e_i V$ for G (because $ge_ig^{-1} = e_{j(g)}$ for each g in G).

In fact, (*) is equivalent to Clifford's Theorem.

2.3. Primitive Linear Groups. We start by stating some facts from the theory of algebras.

Proposition. *Let A be a finite-dimensional simple algebra with centre P.* (1) *If σ is an automorphism fixing P elementwise, there exists an invertible element g of A such that $\sigma(a) = gag^{-1}$ for all a in A.* (2) *If B is a simple subalgebra of A containing the identity of A, then every automorphism τ of B is induced by* *an automorphism* $a \to gag^{-1}$ $(a \in A)$ *of* A, *for suitable* $g \in A$. (3) *The centraliser* C *of the algebra* B *in* A *is a simple subalgebra and* $\dim B \cdot \dim C = \dim A$. *Moreover, the centraliser of* C *in* A *is* B *itself. In particular, the centres of* B *and* C *are the same* (*see* [*Bourbaki* (1958), *Chapter VIII*, § 10]).

We should mention that we are not assuming that the centre of the algebra B in assertion (2) is P.

Theorem (Clifford, 1937). *Let* G *be an irreducible primitive linear subgroup of* $\mathrm{GL}_n(P)$, *and* N *a normal subgroup of* G. *Assume that* P *is algebraically closed. Let* C *be the centraliser of* N *in the matrix ring* $M_n(P)$, *and set* $B = [N]_P$. *The following assertions hold.*

(1) $B \cong M_k(P)$ *and* $C = M_l(P)$ *for some* $k, l \in \mathbb{N}$ *with* $kl = n$.

(2) *There exist subgroups* G_1 *of* $\mathrm{GL}_k(P)$ *and* G_2 *of* $\mathrm{GL}_l(P)$ *contaning the groups* Sc_k *and* Sc_l *of scalar matrices such that*: (a) G *is similar to a subgroup of the Kronecker product of* G_1 *and* G_2; (b) G_1/Sc_k *and* G_2/Sc_l *are homomorphic images of* G; (c) *both* G_1 *and* G_2 *are irreducible and primitive.*

Sketch Proof. It follows from (2.2) that B is simple. By assertion (3) of the proposition, C is also simple. By the Molien-Wedderburn Theorem, $B \cong M_k(P)$ and $C = M_l(P)$ for some k, $l \in \mathbb{N}$, whence it follows that $kl = n$. Since all the irreducible representations of a simple algebra are equivalent, B is similar to the algebra $E_l \otimes M_k(P)$, and its centraliser C is therefore of the form $M_l(P) \otimes E_k$.

For $g \in G$, the map $\beta \to g\beta g^{-1}$, $\beta \in B$, is an algebra automorphism of B. By assertion (2) of the proposition, $g\beta g^{-1} = b\beta b^{-1}$ for all $\beta \in B$ and suitable $b \in B$. Consequently, $b^{-1}g \in C$, so that $g = bc$ for some c in C. The elements b and c are determined uniquely up to multiplication by scalars. If $g_i = b_i c_i \pmod{\mathrm{Sc}_n}$ $(i = 1, 2; g_i \in G, b_i \in B, c_i \in C)$, then $g_1 g_2 = b_1 b_2 c_1 c_2 \pmod{\mathrm{Sc}_n}$. It follows that both of the sets $G_1' = \{b \cdot \mathrm{Sc}_n | g = bc \in G\}$ and $G_2' = \{c \cdot \mathrm{Sc}_n | g = bc \in G\}$ are groups, and the factor-groups G_1'/Sc_n and G_2'/Sc_n are homomorphic images of G. Moreover, it is clear that $G \subset G_1' G_2'$. Transferring the groups G_1' and G_2' over to $M_k(P)$ and $M_l(P)$ respectively, we get assertions (a) and (b). Irreducibility and primitivity of G hold for each larger group, in particular for $G_1' G_2'$. Thus, (c) follows from properties of the Kronecker product (1.2.7).

Corollary A. *Under the assumptions of the theorem, we have*: (a) *if* N *is abelian, it is scalar, and* (b) *if* G *is nilpotent, then* $n = 1$.

Assertion (a) follows from the fact that there are no simple commutative algebras over an algebraically closed field P other than P itself. Item (b) follows from (a), since a nonabelian nilpotent group always has a noncentral abelian normal subgroup.

We can apply the theorem to each of G_1 and G_2. An inductive argument gives:

Corollary B. *Let* G *be a primitive subgroup of* GL_n. *Then there is a decomposition* $n = k_1 \ldots k_r$ *and groups* $G_i \subset \mathrm{GL}_{k_i}(P)$ *such that*: (a) G *is similar to a subgroup of the Kronecker product of* $G_1, \ldots, G_r$; (b) *each* G_i/Sc_{k_i} *is a homomorphic image* *of G*; (c) *each G_i is primitive and every normal subgroup of it is scalar or irreducible*; (d) *the intersection of the kernels of the homomorphisms $G \to G_i/\mathrm{Sc}_{k_i}$ consists of scalar matrices.*

Note that when $r = 2$, the kernel K_1 of the homomorphism $G \to G_1/\mathrm{Sc}_k$ lies in B, and the kernel K_2 of $G \to G_2/Sc_l$ lies in C. Thus $K_1 \cap K_2 \subset B \cap C = P \cdot E_n$. This is the start of the induction for proving (d).

Corollary B gives something of a reduction in the study of primitive groups to groups of a more restricted type, namely groups all of whose normal subgroups are scalar or irreducible. Such a group G is said to be *absolutely primitive*. These groups fall naturally into two classes: those where all soluble normal subgroups are scalar, and those where this is not so. In the first case, additional information for general G is obtained only with difficulty; if G is finite, it has a unique minimal irreducible normal subgroup F; F has a representation as a Kronecker product of copies of a subgroup F_0 of $\mathrm{GL}_r(P)$ for which all proper normal subgroups are central. In this case, n is a power of r; in particular, if n cannot be written as a power m^k with $m, k \in \mathbb{N}$, $k > 1$, then $F = F_0$. To a large extent, the structure of G is determined by that of F_0.

The second case – that where G contains a nonscalar soluble normal subgroup – can be studied further. Let A be a minimal nonscalar soluble subgroup of G. Then its commutator subgroup A' is scalar. Moreover, $|A'| = p$ for some prime p, and A/A' is an elementary abelian p-group of order n^2. In particular, $n = p^m$ and $|A/A'| = p^{2m}$ for some $m \in \mathbb{N}$.

We shall view A/A' as a vector space over the field $\mathbb{Z}_p$ of p elements. The operation $(a, b) \to aba^{-1}b^{-1}$, for a, $b \in A$, of commutation induces a nondegenerate skewsymmetric bilinear form on A/A', while the automorphisms $a \to gag^{-1}$ ($a \in A$, $g \in G$) induce linear maps of A/A' preserving this form. At the same time, we get a homomorphism of G/A into the symplectic group $\mathrm{Sp}_{2m}(\mathbb{Z}_p)$ (Hall and Higman 1956, D.A. Suprunenko (1972)). If G is absolutely primitive, the subgroup of $\mathrm{Sp}_{2m}(\mathbb{Z}_p)$ thus obtained is irreducible. If $p \neq 2$, or $p = 2$ but A contains a matrix iE_n with $i^2 = -1$, $i \in P$, and M denotes the normalizer of A in $\mathrm{GL}_n(P)$, then $M/A \cdot \mathrm{Sc}_n(P) \cong \mathrm{Sp}_{2m}(\mathbb{Z}_p)$.

2.4. Techniques from Field Theory. Generally speaking, the structure of a linear group depends critically on the ground field. The principal types of fields over which linear groups are studied are the algebraically closed fields, finite fields, fields of algebraic numbers and fields with valuation (including $\mathbb{R}$). Another class of fields of a very general nature that play a significant role are the finitely generated fields. This is because a finitely generated linear group can be written over a finitely generated field. A typical example of the use of field theory techniques is the proof of the next theorem.

Theorem. *Let G be a finitely generated subgroup of* $\mathrm{GL}_n(P)$. *Then there exists a number $m = m(G)$ such that $g^m = 1$ for every element g of finite order in G.*

Proof. We may assume that P is algebraically closed. Let P_G be the least subfield of P containing the entries of the matrices generating G; then

$G \subset \mathrm{GL}_n(P_G)$. The field P_G is finitely generated. Let P_0 be the prime subfield of P_G (= the smallest subfield of P_G). As is known, for every finitely generated field F there exists a number $r = r(F)$ such that if λ is an element of P which is a root of a polynomial f_1 with coefficients in P_0 and also a root of some other polynomial f_2 of degree t with coefficients in F, then λ is a root of a polynomial f_3 of degree at most rt having coefficients in P_0. Let g be an element of finite order in G. For λ we take an eigenvalue of g, for f_2 its characteristic polynomial, and $f_1 = x^s - 1$, where s is the order of g in G. Then λ is a root of unity satisfying an equation of degree at most rn with coefficients in P_0. It is a result from algebraic number theory that $\lambda^q = 1$ for some q depending on r an n. Thus, q depends on P_G rather than on G. Since λ is any eigenvalue, all eigenvalues of g^q are 1. On reducing g^q to Jordan normal form over P, we see that it is E if the characteristic p of P is 0, while for $p > 0$ its p^n-th power is E. Thus, we can set $m = q$ for $p = 0$, and $m = qp^n$ for $p > 0$.

Corollary A (Schur 1911). *Finitely generated periodic linear groups are finite.*

We recall that a group is *periodic* if all its elements are of finite order.

Corollary B. *Let P be an algebraically closed field of characteristic $p > 0$. If the group G in the statement of the theorem is infinite, then G contains a diagonalisable matrix g such that $g^r \neq E$ for all r in $\mathbb{N}$.*

For each element g of G, we write $g = du$, where $du = ud$, d is diagonalisable and u is unipotent (this is the Jordan decomposition). Since $p > 0$, $u^{p^n} = E$ and thus $g^{p^n} = d^{p^n}$ is a diagonalisable matrix in G. If g^{p^n} is of finite order for each g in G, the theorem tells us that $g^m = E$ for some m. But G is therefore finite, by Corollary *A*.

§ 3. Topological Methods in the Theory of Linear Groups

Topological methods occupy a significant position in linear group theory. Applications are associated with three types of topology: the Euclidean topology, the topology of a normed space over a field with non-Archimedean valuation, and the Zariski topology. In the first of these, metric properties of the relevant spaces are used in addition to topological properties.

3.1. The Hermitian Metric. The use of properties of the Hermitian metric in analysing abstract linear groups was initiated in 1878 by Jordan. He used this method to prove the following theorem, which remains to this day one of the deepest results in linear group theory.

Theorem A. *Let G be a finite subgroup of $\mathrm{GL}_n(\mathbb{C})$. Then G has an abelian normal subgroup of index not more than a certain quantity $J(n)$ depending only on n.*

As we know from 1.3.13, G is similar to a subgroup of the group $U_n(\mathbb{C})$ of unitary matrices. Thus, we may assume that $G \subset U_n(\mathbb{C})$. A decisive point in the proof is to see that a group H generated by the matrices g in G such that $\|g - E\| < 1/2$ is abelian. The norm $\|\cdot\|$ on $M_n(\mathbb{C})$ is that introduced in 1.4.2. We have $\|ab\| = \|a\| = \|b\|$ for all $a, b \in U_n(\mathbb{C})$. If $g_1, \ldots, g_r$ are coset representatives of G modulo H, then $\|g_i - g_j\| \geqslant 1/2$ for $i \neq j$. When a complex matrix is represented as a point in $2n^2$-dimensional real space, the elements $g_1, \ldots, g_n$ can be thought of as points on the surface of the unit sphere such that the distance between any two of them is at least 1/2. It is clear that the number of points like this is finite, and that it is bounded in terms of n. The proof that H is abelian is based essentially on the following curious assertion about arbitrary unitary matrices a, b: if a and bab^{-1} commute but $ab \neq ba$, then $\|E - b\| \geqslant 4$. The reader can refer to [Curtis and Reiner (1962), Chapter 5, § 36] for more details. Note that $J(n)$ is often referred to as the Jordan function.

Jordan's successors made very skilful use of properties of the Hermitian metric in proving a wide class of very diverse theorems about finite linear groups. Some examples of this are to be found in Blichfeldt's book (1917); see also Robinson (1989).

It is completely natural that topological techniques should be used to investigate topological linear groups, including Lie groups. However, it is not part of our purpose here to expound this sort of question. Discrete matrix groups over $\mathbb{C}$ and $\mathbb{R}$ are developed within the confines of the theory of Lie groups; see Ragunathan (1972), for example. However, certain aspects of the theory of discrete transformation groups, for example discrete groups of affine transformations generated by reflections are developed without using the theory of Lie groups (see Bourbaki (1968) Chapter 5]). We note that the theory of discrete transformation groups of (high-dimensional) Lobachevskij space has been much developed of late; this is particularly so of groups generated by reflections in hyperplanes.

3.2. Linear Groups over Locally Compact Fields with Non-Archimedean Valuations. The techniques outlined in 3.1 are highly traditional in linear group theory: topological arguments in Euclidean and Hermitian spaces are much used, and for very different reasons. Thanks to Tits's work on free subgroups of linear groups (see § 4 below), it has become clear that this sort of method, developed as it was for matrices over fields with non-Archimedean valuations, can also be very useful in the theory of linear groups. The most significant part of this work of Tits establishes deep connections between algebraic, geometric and topological properties of linear groups over locally compact fields. As for the purely topological machinery, its development is in large measure analogous to that of finite-dimensional normed spaces over $\mathbb{R}$ and $\mathbb{C}$. One of the significant events here is the introduction of a metric on projective space, whereupon a very natural connection appears between the norms of the eigenvalues of a linear map and topological properties of its action. Let us explain in more detail.

We recall that two metrics d_1 and d_2 on a set M are said to be *equivalent* if there exist constants c_1, c_2 such that $c_1 d_2(x, y) \leq d_1(x, y) \leqslant c_2 d_2(x, y)$ for all

$x, y \in M$. Let V be a finite-dimensional linear space over a field P with valuation v. If $B = (b_1, \dots, b_n)$ is a basis and $x = x_1 b_1 + \cdots + x_n b_n$, $y = y_1 b_1 + \cdots + y_n b_n$ are two vectors, we set

$$\mu_B(x, y) = \sup_{1 \leqslant i \leqslant n} v(x_i - y_i).$$

It is easy to check that the metrics μ_B are mutually equivalent.

Now let $\bar{V}$ be the projective space associated with V. If $(x_1, \dots, x_n)$ is the coordinate system in V corresponding to the basis B, the equation $x_1 = 0$ gives a hyperplane L in V and the set $(x_1^{-1}x_2, \dots, x_1^{-1}x_n)$ can be regarded as coordinates on $\bar{V} - L$. In this sense, we shall say that it is the system of affine coordinates on V corresponding to the basis B, with domain of definition $D_B = \bar{V} - L$. For $x, y \in D_B$, we set $d_B(x, y) = \sup_{i>1} v(x_1^{-1}x_i - y_1^{-1}y_i)$. It is not hard to check that d_B is a metric on D_B; moreover, if B, C are bases in V and K a compact subset of $D_B \cap D_C$, the metrics d_B and d_C are equivalent on K.

A metric d on $\bar{V}$ is said to be *admissible* if it is consonant with the topology on $\bar{V}$ and if, for every basis B of V and every compact subset K of D_B, the metrics d and d_B are equivalent on K. If P is locally compact, there always exists an admissible metric on $\bar{V}$. Moreover, every projective transformation A of $\bar{V}$ has finite norm $\|A\|$. Here, as usual, for an arbitrary set $M \subset \bar{V}$ (in particular, for $M = \bar{V}$), the norm $\|A\|_M$ is defined by the following formula:

$$\sup_{x, y \in M, x \neq y} d(Ax, Ay)/d(x, y),$$

provided that M has more than one point; otherwise we set $\|A\|_M = 0$. When $M = \bar{V}$, we write $\|A\|$ instead of $\|A\|_{\bar{V}}$.

Let g be a semisimple element of $\mathrm{GL}(V)$, and $\lambda_1, \dots, \lambda_n$ its eigenvalues. It makes sense to speak of the quantities $v(\lambda_i)$ since the valuation v has a unique extension to any given finite extension of P (see 1.4.4; note that every locally compact field is complete in its valuation). As is well known, in the case of an algebraically closed field, there is a spectral decomposition of V as a direct sum of subspaces, on each of which g acts like a scalar. We have the following curious fact in the general case: V is a direct sum of g-modules, on each of which the eigenvalues of g have identical valuations. These eigenvalues do not necessarily lie in P. We denote by V_g the g-module corresponding to the eigenvalues with greatest valuation, and by V^g the sum of all the other such g-modules.

As with the complex field, when $v(\lambda_i) < 1$ the elements g^m $(m = 1, 2, \dots)$ shrink V to the point 0; more exactly, for every compact subset K of V and every neighbourhood U of 0 there exists a number $m \in \mathbb{N}$ such that $g^m K \subset U$. There is a distinctive analogue of this result for projective space. Suppose that $\dim V_g = 1$. Then $\bar{V}_g$ is a point in $\bar{V}$ and every compact subset K of $\bar{V}$ disjoint from $\bar{V}^g$ shrinks to it. Moreover, there is the following version of the "contraction mapping" principle. Suppose that $K \subset \bar{V}$ and $K \cap \bar{V}_g = \varnothing$, and let K^0 be the interior of K. If $g^m K \subset K^0$ and $\|g^m\|_K < 1$ for some $m \in \mathbb{N}$, we have $\dim V_g = 1$; moreover, the sets $g^{ml}K$ $(l = 1, 2, \dots)$ shrink to the point $\bar{V}_g$ in K^0.

We conclude this section with a fairly easy argument that explains the role of this technique in the proof of Tits's theorem about free subgroups.

Set $\Omega_G = \{x \in G | \dim V_x = \dim V_{x^{-1}} = 1\}$, and $\overline{V}_{x*} = \overline{V}_x \cup \overline{V}_{x^{-1}}$.

Theorem. *Let Y be a finite subset of Ω_G consisting of semisimple elements, and suppose that $\overline{V}_{x*} \subset \overline{V} - \overline{V}_{y*}$ for any different elements x, y of Y. Then, there exists a number r in $\mathbb{N}$ such that the group generated by the elements y^r ($y \in Y$) is free.*

Proof. Choose any point w in $\overline{V} - \bigcup_{y \in Y} \overline{V}_{y*}$. Clearly, for every y in V there exist compact neighbourhoods U_y and $U_{y^{-1}}$ of $\overline{V}_y$ and $\overline{V}_{y^{-1}}$ respectively such that $U_y \cup U_{y^{-1}} \subset \overline{V} - \bigcup_{x \in Y, x \neq y} V_{x*}$. Set $U_{y*} = U_y \cup U_{y^{-1}}$. Then, as remarked above, there exists a number r in $\mathbb{N}$ such that $y^m(U_{y*} \cup w) \subset U_{y*}$ when $|m| \geqslant r$, for any different elements x, y of Y. It follows that the group generated by the elements y^r is free, since for every sequence $t_1, \ldots, t_s$ of elements of $Y \cup Y^{-1}$ we have $t_1^{rl_1} t_2^{rl_2} \ldots t_s^{rl_s}(w) \subset U_{t_1^*}$; in particular, $t_1^{rl_1} \ldots t_s^{rl_s} w \neq w$, so that $t_1^{rl_1} \ldots t_s^{rl_s} \neq 1$.

3.3. Zariski Topology Techniques. The Zariski topology is a very effective means of proving widely different theorems about linear groups. The principal element in its use is induction on the dimension of the linear group, in the sense of 1.5.5. (We stress that we are talking about the induced topology: see 1.5.4). This dimension is a significantly finer invariant than, say, the matrix degree or the dimension (*qua* vector space) of the enveloping algebra $[G]_P$ of G. Zariski topology methods constitute what is, in essence, an elementary part of a universal method in algebraic groups: we shall retrun to this theme in § 4.

The following result of Platonov (1965; see also Platonov (1966), Wehrfritz (1973)) is an example where induction on the dimension – combined with Chevalley's Theorem (1.5.4) – is of good effect.

Theorem. *The Sylow p-subgroups of a periodic linear group are conjugate.*

We recall that a Sylow p-subgroup is a p-subgroup that is maximal under the relation of inclusion.

Proof. Let G be a periodic subgroup of $\mathrm{GL}_n(P)$. We proceed by induction on $\dim G$. If $\dim G = 0$, we simply quote Sylow's Theorem for finite groups (since the index of G^0 in G is finite: see 1.5.5). It is easy to see that conjugacy of Sylow p-subgroups is preserved in finite extensions of locally finite groups where they are conjugate. Therefore, we may assume that G is connected. Let S_1 and S_2 be two Sylow p-subgroups of G. Clearly, there exist finite subsets $H_i \subset S_i$ such that $S_i \subset [H_i]_P$, where $[H_i]_P$ is the linear P-space spanned by H_i ($i = 1, 2$). By Corollary A in 2.4, $\langle H_1, H_2 \rangle = H$ is finite. By Sylow's Theorem, on replacing S_2 by hS_2h^{-1} for suitable h in H, we may assume that H is a p-group. Set $G_1 = C_G(Z(H))$, where $Z(H)$ is the centre of H. Then $G_1 \supset \{S_1 \cup S_2\}$. Since G_1 is closed (see (1.5.4(a)), by the induction hypothesis it is enough to consider the case where $\dim G_1 = \dim G$; and this is possible only when $G_1 = G$. Thus, $Z(G) \neq 1$. It is easy to see that the Sylow p-subgroups of a periodic group are conjugate if they are conjugate in the central factor-group. Since $Z(G)$ is closed, there exists a continuous homomorphism $\phi: G \to \mathrm{GL}_m(P)$ with kernel $Z(G)$, for some $m \in \mathbb{N}$ (see 1.5.4). If $Z(G)$ is infinite, we have $\dim \phi(G) < \dim G$, and we can use induction. Suppose that $|Z(G)| < \infty$. By repeating the argument with G replaced by $\phi(G)$, we may assume that $Z(\phi(G)) \neq 1$. Let Z_1 be the inverse image of $Z(\phi(G))$ in G. It is clear that the map $\tau: g \to gag^{-1}a^{-1} \in Z(G)$ $(g \in G, a \in Z_1 - Z(G))$ is a group homomorphism, and that $k = |G\colon \operatorname{Ker} \tau| < \infty$. Since G is connected, $k = 1$ (see 1.5.4), so that $a \in Z(G)$; this is a contradiction.

Kaplansky's small book (1957) contains a proof of the following theorem of Kolchin (1948) based on Zariski topology techniques.

Theorem B. *Every connected soluble matrix group over an algebraically closed field is triangularisable* (*that is, is similar to a subgroup of some* $T^n(P)$).

Every linear group has a connected normal subgroup of finite index. Thus, from Theorem B we get:

Theorem C. *Every soluble matrix group over an algebraically closed field has a triangularisable normal subgroup of finite index.*

The following general result is obtained by going over to the algebraic closure of the ground field.

Theorem D (Kolchin, Mal'tsev). *Every soluble linear group has a normal subgroup of finite index with unipotent commutator subgroup.*

The next theorem, due to Chevalley, is of a somewhat different nature. It plays an important part in the solution of problems in recognising linear groups (see § 5).

Theorem E. *Let P be an algebraically closed field. If $H_1, \ldots, H_m$ are closed connected subsets of* $\mathrm{GL}_n(P)$ *such that $E_n \in H_i$ for $i = 1, \ldots, m$, then the subgroup $\langle H_1, \ldots, H_n \rangle$ is also closed and connected.*

For the proof, see [Humphreys (1975), Chapter II, § 7.5].

A more systematic exposition of the applications of the Zariski topology in abstract linear groups is to be found in Wehrfritz's book (1973). See also Dixon (1971), Zalesskij (1981).

3.4. Fundamental Regions of Transformation Groups and Presentations of Linear Groups. There is an effective and well-developed method for finding systems of generators and relations (that is, presentations) for many different classes of groups. To simplify somewhat, it consists in realizing the group as a group of transformations of a suitable topological space having a fundamental region, and describing the group in terms of its action in a neighbourhood of the fundamental region. This method is applicable to all groups; however, discovery of a successful realisation of an abstract group as a transformation group requires great skill. At the same time, such realisations occur naturally for many important classes of groups, or at the very least are natural in character. A more detailed treatment of this method requires much more on the theory of transformation groups, and this could lead us beyond our main theme. For this reason, we shall restrict ourselves to giving a certain amount of explanation, and to covering a number of typical results. The reader will then be able to get some idea of what has been achieved and what is possible. A systematic exposition of this subject is to be found in the survey by Vinberg and Shvartsman (1988).

Let M be a topological space on which some group G acts by continuous transformations. A *fundamental region* for G in M is a subset F such that $GF = M$ and different points of F lie in different G-orbits. Thus, a fundamental region consists of a set of representatives of the orbits of G on M, one from each orbit. The group G acts naturally on the set $T = \{gF\}_{g \in G}$, the action being *simply transitive* (that is, $g = 1$ if $gF = F$, and indeed $g = 1$ if $ghF = hF$ for an arbitrary $h \in G$). Often, T carries a natural G-invariant reflexive relation of proximity of elements, or of elements being contiguous. For example, we could declare that gF and hF are contiguous if $g\bar{F} \cap h\bar{F} \neq \varnothing$. It is clear that gF and hF are contiguous if and only if F and $g^{-1}Fg$ are. Therefore, the set $G_F = \{g \in G \mid F \text{ and } gF \text{ are contiguous}\}$ determines this relation. The choice of the set F and the relation of contiguity can be viewed as successful if it is possible to get from F to gF (for arbitrary g in G) by passing from neighbour to neighbour, that is, if there exists a finite sequence $g_1, g_2, \ldots, g_n$ such that g_iF and $g_{i+1}F$ are contiguous for $i = 1, \ldots, n-1$, that is, $g_{i+1}^{-1}g_i \in G_F$. Clearly, G_F generates G in this case. In a range of typical examples, G_F is finite. If we can get from F to G_F by two different routes $\{g_1, \ldots, g_n\}$ and $\{h_1, \ldots, h_m\}$ (here $g_i, h_i \in G_F$), the element $g_1 \ldots g_n h_m^{-1} \ldots h_1^{-1}$ is a relation between the elements of G_F. It can be counted as an essential relation if no proper subsequence of $g_1 \ldots h_1^{-1}$ is a relation. It is clear that every relation between elements of G_F reduces to an essential relation as described above.

The contiguity relation can be expressed conveniently in terms of a graph. The vertex set is T, and vertices gF and hF are joined by an edge if and only if gF and hF are contiguous. The graph thus obtained is non-oriented and has no loops nor multiple edges, and G acts transitively on the vertices. Then G_F corresponds to the set of edges containing F, and the set of essential relations to the minimal cycles involving the vertex F. Moreover, the inverse of every element in G_F is also contained in G_F, and the relation $gg^{-1} = 1$ corresponds to the "trivial cycle', moving from F to gF and back. Of course, we can also consider the corresponding oriented graph. Note that it is enough to traverse a cycle in one direction, because the inverse path gives a derivative relation. The reader will find a fairly accessible treatment of this technique in the book of Coxeter and Moser (1972). As an example, we mention the group of isometries of the plane generated by reflections s_1, s_2, s_3 in the sides of an equilateral triangle. In this case, the plane is divided into three series of triangles each arranged in parallel lines. If we declare triangles having a common edge to be contiguous, then $G_F = \{s_1, s_2, s_3\}$, and minimal cycles take the form $F, s_1F, s_1s_2F, s_1s_2s_1F, s_1s_2s_1s_2F, s_1s_2s_1s_2s_1F, s_1s_2s_1s_2s_1s_2F = F$ (up to the ordering of the s_i): this is the path through contiguous triangles around a single vertex. Together with the trivial cycle $F, s_iF, s_is_iF = F$, this gives the following set of defining relations for the group: $(s_is_j)^3 = 1, s_i^2 = 1$ $(i, j = 1, 2, 3)$. However, the relations $(s_is_j)^3 = 1$ and $(s_js_i)^3 = 1$ are dependent relations obtained by traversing the same cycles in opposite directions. Observe that this naïve example represents an extensive circle of like situations. They arise in cases when a group G acts as a group of isometries of n-dimensional space (or a sphere, or any space of constant curvature) in such a way that the closure of a fundamental region is a polyhedron (or a geodesic polyhedron). In this case, for G_F we take the set of all elements g of G such that F and gF have a common $(n-1)$-face. The minimal relations arise when we traverse mutually contiguous regions gF having common $(n-2)$-faces.

Topological properties of the region F, such as connectedness, convexity *etc* play an essential part in the choice of G_F and the discovery of minimal relations. Problems of determining fundamental regions arise in the theory of discrete subgroups of Lie groups and the arithmetic theory of algebraic groups. Under certain definite restrictions, finite generation or finite relatedness of the corresponding groups are thereby established. The classical result here is the theorem of Borel and Harish-Chandra (see 1.6.11). We cannot dwell on this subject in more detail, and refer the reader to more specialist texts. Ragunathan's book (1972) is devoted to the study of discrete subgroups of Lie groups. A reasonably complete exposition of the arithmetic theory of algebraic groups is to be found in the book of Platonov and Rapinchuk (1991); Humphreys's book (1980) may serve as an introduction.

Many results about generators and relations for linear groups arise because one can make them act on discrete metric spaces and then study their fundamental regions. The provenance of the spaces can be exotic; in the typical example of the general linear group over a local field, it is the set of lattices relative to the general linear group over the ring of integers of the given field (Behr *et al*, 1962). Later on, the connection between results of this sort and the theory of groups with BN-pairs was elucidated. A group with a *BN-pair* (or *Tits system*) is a group with a distinguished pair B, N of subgroups satisfying certain conditions (see Bourbaki (1968), Chapter 4, §2.1)]. The main features of the theory of groups with BN-pairs were developed by Tits in 1974. It is an axiomatisation of the algebraic group property of having a Bruhat decomposition. In the classical case, B is a Borel subgroup and N the normaliser of a maximal torus (see 1.6.2). We shall be interested in another version of the concept of Bruhat decomposition, discovered by Iwahori and Matsumoto in 1965. For G we consider an algebraic group over a local field, while B is its Iwahori subgroup. In the case of the general (or special) linear group, the Iwahori subgroup is constructed like this. Let K be the valuation ring and I the valuation ideal. The *Iwahori subgroup* is the set of all matrices over K with elements of I standing below the main diagonal. If N is the normaliser in G of the group of diagonal matrices and W a transversal of N modulo $N \cap B$, the Bruhat-Iwahori decomposition is $G = \bigcup_{w \in N} BwB$. The pair B, N then satisfy all the conditions in the definition of group with a BN-pair.

We recall further that a *parabolic* subgroup of a group G with a BN-pair is a subgroup containing some conjugate gBg^{-1}, $g \in G$. Clearly, G acts by conjugation on the set of all parabolic subgroups. Under certain conditions (which are automatically satisfied if the valuation on P is discrete and the BN-pair is constructed from an Iwahori subgroup), analysis of a fundamental region of this action produces a representation of G as a free product with amalgamation of maximal parabolic subgroups containing B (Tits: see [Serre (1977), Chapter II]). The following theorem of Tits holds for absolutely arbitrary groups with BN-pairs.

Let $\{G_\alpha\}$ be the set of minimal parabolic subgroups such that $G_\alpha \supsetneqq B$. Then G is the free product of N and the G_α with the intersections amalgamated. A proof is provided in the lectures of Serre [(1977), Theorem 8]. There are versions of this theorem having to do with decreasing the number of constitutent subgroups (suitable replacements for the G_α are larger parabolic subgroups); the reader will find examples in Serre (1977). Perhaps the most important of these is the decomposition of G as a free product with amalgamation of standard parabolic subgroups of rank 2, that is, subgroups containing some G_α as maximal subgroups. An important question is to determine the subgroups H of G for which there is a decomposition as in Tits's theorem with G_α replaced by $H \cap G_\alpha$. The special case where the set $\{G_\alpha\}$ has exactly two members is investigated in detail in Serre's lectures (1977). Roughly speaking, a suitable decomposition exists in this case for practically every infinite subgroup H (see [Serre (1977), Chapter II, Theorem 3]). This fact emerges as a result of the development of a general theory of groups acting on graphs of a special kind called trees (a tree is a connected graph without cycles).

§4. Algebraic Group Methods

The essence of this method is to embed the given linear group $G \subset \mathrm{GL}_n(P)$ in an algebraic group $\bar{G}$, and then to use the structure theory of algebraic groups. Normally, $\bar{G}$ is the closure of G in the Zariski topology. It is preferable to take P to be algebraically closed, since this is the case where the structure theory is most decisive. The wide spectrum of questions that can be solved using the algebraic group methods makes it the fundamental tool for investigating infinite linear groups.

The proof of the following assertion is the simplest example where the structure theory of algebraic groups is applied.

Theorem A (Platonov 1967). *A linear group satisfies an identical relation if and only if it is a finite extension of a soluble group* (*or even of a triangularisable group if the ground field is closed*).

The sufficiency is completely obvious. To establish necessity, we embed the given group G in a group of matrices over an algebraically closed field P containing an element t transcendental over the prime field. Let $\bar{G}$ be the closure of G in the Zariski topology of P. Note then that an identical relation $X_{i_1}^{\mp 1} X_{i_2}^{\mp 1} \dots X_{i_k}^{\mp 1} = 1$ is equivalent to a system of n^2 polynomial equations in the matrix elements. Thus, G has an identical relation of this sort if and only if $\bar{G}$ satisfies the same identity. It is therefore enough to show that $\bar{G}$ is a finite extension of a soluble group. Using general facts from the structure theory (see

1.6.6), everything comes down to the case where G is connected and semisimple. The decisive step is to use the fact (1.6.5) that a nontrivial semisimple group contains a subgroup isomorphic to $\mathrm{SL}_2(P)$ or $\mathrm{PSL}_2(P)$. Now just observe that these groups cannot satisfy an identical relation, since they contain free subgroups (for example, $\mathrm{SL}_2(P)$ contains the free subgroup generated by $\begin{pmatrix} 1 & 0 \\ t & 1 \end{pmatrix}$ and $\begin{pmatrix} 1 & t \\ 0 & 1 \end{pmatrix}$).

Therefore, if G does not satisfy an identical relation, its closure in $\mathrm{GL}_n(P)$ contains a free subgroup. Apparently, this circumstance has caused certain mathematicians to ask whether G itself must contain a free subgroup. It has proved necessary to suppose that $\operatorname{char} P = 0$, because of the obvious counterexample when $p > 0$ provided by the group $\mathrm{GL}_n(P_a)$, where $n > 1$ and P_a is the algebraic closure of the prime subfield P_0 of P. The group $\mathrm{GL}_n(P_a)$ is locally finite and insoluble. This obstacle does not arise if G is finitely generated. The answer was found by Tits in 1972:

Theorem B. *Every finitely generated linear group has a nonabelian free subgroup or is soluble-by-finite.*

It follows that a linear group has a nonabelian free subgroup or else is soluble-by-locally finite, while in the characteristic zero case the theorem holds without the assumption about finite generation.

Note that Theorem A follows from Theorem B.

The proof of Tits's theorem uses the following result, which is of independent interest.

Theorem C. *Let G be an infinite absolutely irreducible linear group on finitely many generators. Then G contains a semisimple element of infinite order.*

This follows immediately from assertion (E) of 1.6.5. For, let $\overline{G}$ be the closure of G in the Zariski topology over an algebraically closed field; then $\overline{G}$ and its connected component $\overline{G}^0$ are reductive. By 1.6.5, the set of non-semisimple elements of $\overline{G}^0$ is contained in a proper closed subset M. This is true *a fortiori* for G^0 – its non-semisimple elements lie in $M_1 = M \cap G$. The complement $G^0 - M_1$ consists of semisimple elements. The set F of elements f of finite order in G satisfies the relation $f^r = 1$ for some (fixed) r (see 2.4); thus, this set is closed. It follows that the union $M_2 = (F \cap G^0) \cup M_1$ is closed. Its complement $R = G^0 - M_2$ is infinite, otherwise the connected set G^0 would be the union of finitely many closed subsets. But R consists precisely of the semisimple elements of infinite order in G^0. Thus, Theorem C really asserts the existence of infinitely many semisimple elements of infinite order (more than that, the closure of the set of all such elements contains G^0). There is a refinement of Theorem C concerning the existence in G^0 of semisimple elements of infinite order with the additional property that their centralisers in G^0 (and even in $\overline{G}^0$) contain no unipotent elements.

Here is a rough plan of the proof of Theorem B. Firstly, the question reduces to the case where the ground field is locally compact, the group G is connected and contains a semisimple element having an eigenvalue of multiplicity 1 with norm strictly greater than that of all its other eigenvalues. A central idea in the proof is to establish the fact that, with these assumptions, G has a dense subset consisting of semisimple elements with a unique eigenvalue of multiplicity 1 which is maximal in norm, and a unique eigenvalue minimal in norm. After all that, the theorem from section 3.2 is used to find a free subgroup.

§ 5. Residual Methods

The residual method is an effective tool for investigating finitely generated linear groups, and, using the knowledge acquired, for investigating arbitrary linear groups. The idea is due to Mal'tsev, who proved the following theorem in 1940.

Theorem A. *Let G be a finitely generated subgroup of $\mathrm{GL}_n(P)$. For every finite set $h_1, \ldots, h_r$ of elements of G, there exists a homomorphism ϕ: $G \to \mathrm{GL}_n(F)$, where F is a finite field and $\phi(h_i) \neq 1$ for $i = 1, \ldots, r$. In other words, G is residually finite linear of degree n.*

Mal'tsev used this theorem to solve the Hopficity problem in the class of finitely generated linear groups: such groups are not isomorphic to proper factor-groups of themselves. The Mal'tsev approach is essentially algebraic-geometrical in nature: he observed that every finitely generated matrix group $G \subset \mathrm{GL}_n(P)$ can be associated with an algebraic variety defining G up to isomorphism. In more detail: we supplement the generating set $a_1, \ldots, a_m$ of G with elements $b_1, \ldots, b_m$ such that $a_i b_i = 1$ for $i = 1, \ldots, m$. Defining relations for G can then be written as a system of polynomial equations in the entries of the matrices $a_1, b_1, \ldots, a_m, b_m$, with coefficients in $\mathbb{Z}$ if char $P = 0$ and in $\mathbb{Z}_p$ if char $P = p > 0$. On replacing the matrix entries $(a_i)_{kl}, (b_j)_{kl}$ by unknowns x_{ikl}, y_{jkl}, we get a system of polynomial equations in $2mn^2$ unknowns, and the matrix entries $(a_i)_{kl}, (b_j)_{kl}$ give a solution of this system. Therefore, the generating set for G is a point in some Zariski-closed set M in a $2mn^2$-dimensional vector space W (see 1.5.2). Since a set of matrices is a point of M precisely when they satisfy all the defining relations of G, it follows that every point gives a group in which all relations of G hold, that is, a homomorphic image of G. Since the coefficients of the equations defining M lie in $\mathbb{Z}$ (in $\mathbb{Z}_p$ respectively), if desired the solution of these equations can be written not only over P but also over any field L (field L of characteristic p, respectively). If L is of positive characteristic q, we replace the equations for M by a system of equations over $\mathbb{Z}_q$ by taking all the coefficients modulo q. Denote the resulting set of equations by M_L. At this point, Mal'tsev used two assertions from algebraic geometry: (a) if $M \neq \varnothing$ and $p = 0$, then $M_{\mathbb{Z}_q} \neq \varnothing$ for infinitely many primes q; (b) if $M \neq \varnothing$ and $p > 0$, then $M_L \neq \varnothing$ for a suitable finite field L of characteristic p. Moreover, these assertions continue to hold if instead of M we take the set M' obtained from M by removing the solutions of any finite system of polynomial equations in x_{ikl} and y_{jkl}. This last system arises in Mal'tsev's arguments out of the requirement that none of the elements $h_1, \ldots, h_r$ can be the identity matrix. This proves the existence of the desired homomorphism ϕ, and completes our survey of Mal'tsev's idea.

Platonov's approach is more constructive. The starting-point is the following fact from commutative algebra:

(*) Every finitely generated commutative ring with 1 having no zero divisors is residually a finite field. In other words, the intersection of the maximal ideals of such a ring is the zero subring.

To be more precise, if the characteristic of the ring in (*) is 0, the characteristics of the finite fields arising run over almost all primes.

Mal'tsev's residual theorem follows immediately from (*): every finitely generated subgroup G of $\mathrm{GL}_n(P)$ is contained in $\mathrm{GL}_n(R)$, where R is the subring generated by the matrix entries of the generators for G. It is clear that R is finitely generated, so that $\mathrm{GL}_n(R)$ is residually in the set of finite groups of the form $\mathrm{GL}_n(R/R_i)$, where the R_i run over the maximal ideals of R. In particular, G is residually in a set of subgroups G_i of the $\mathrm{GL}_n(R/R_i)$. However, Platonov's approach enables one to discern far more of the detail associated with the structure of the groups giving the residual properties. These are finite, and knowledge of their arithmetic invariants turns out to be very important in the process of determining properties of G itself.

Theorem B (Platonov 1966). *Let R be a finitely generated ring of characteristic q with no zero divisors, and let $p_1, \ldots, p_m$ be primes different from q. There is a set $\{R_i\}$, $i = 1, 2, \ldots,$ of maximal ideals of R such that $\bigcap R_i = 0$ and the orders of the Sylow p_j-subgroups of $\mathrm{GL}_n(R/R_i)$ are bounded by a number not depending on i nor on j.*

This theorem yields at once the following assertion, which is a strengthened form of the Mal'tsev residual theorem.

Theorem C. *Let G be a finitely generated subgroup of $\mathrm{GL}_n(P)$ and $p_1, \ldots, p_m$ finitely many primes different from* char P. *Then, G is residually in a set of finite groups G_i such that:*

a) *for $j = 1, \ldots, m$, the orders of the Sylow p_j-subgroups of the G_i are bounded by a number independent of i and j;*

b) *G_i has a faithful representation by $n \times n$ matrices over a finite field of characteristic different from $p_1, \ldots, p_m$.*

The nilpotency of the Frattini subgroups of finitely generated linear groups can be deduced from a). We recall that the *Frattini subgroup* is the intersection of the maximal subgroups.

It is clear from all this that Platonov's method of establishing residual properties is governed principally by the choice of the ideals R_i of R with finite quotients R/R_i. The following theorem produces very different residual properties.

Theorem D (Platonov 1968). *Let G be a finitely generated linear group, $G \subset \mathrm{GL}_n(P)$.*

(a) *If* char $P = p > 0$, *G has a normal subgroup G_1 of finite index that is residually a finite p-group.*

(b) *If* char $P = 0$, *then G has a normal subgroup G_1 of finite index that is residually a finite p-group for all but finitely many primes p, the exceptions depending on G.*

Corollary (Selberg). *Under the hypothesis of item* (b) *of Theorem D, G_1 has no nontrivial elements of finite order.*

This follows from the fact that a group must be torsionfree if it is residually finite-p and residually finite-q for two different primes p and q. An important refinement of (b) in Theorem D is obtained by Lubotzky (see Theorem D in 3.1.2 below).

Unquestionably, the fact that the classification theorem for finite simple groups is complete gives many new possibilities for the residual method.

Let H be the closure of a group G in $\mathrm{GL}_n(P)$ in the Zariski topology; here $\bar{P}$ is an algebraically closed field containing P. Some recent papers (Matthews (1982), Weisfeiler (1984) *et al*) investigate the connection between H and the groups G_i in the argument following (*) above. In particular, they study the case where H is a simple group defined over $\mathbb{Q}$ and realised in its natural action on its Lie algebra. It turns out in this case that, for almost all i, G_i is just the group H_R/H_{R_i} of H_R/H_{R_i}-points of H over R/R_i. (The notation H_{R/R_i} does not always make sense; however, we explained at the beginning of this section that all is well if H is given by equations with coefficients in $\mathbb{Z}$.) For example, if H is $\mathrm{SL}_n(\mathbb{C})$ in its adjoint representation and G is a subgroup of $\mathrm{SL}_n(\mathbb{Z})$ dense in $\mathrm{SL}_n(\mathbb{C})$ with respect to the Zariski topology in $\mathbb{C}$, then G_i is almost always isomorphic to $\mathrm{SL}_n(R/R_i)$.

§ 6. Recognising Linear Groups

In practice, it often happens that we have only incomplete information about a linear group. The situation would be greatly improved if it could be identified as some familiar group. Together with various general considerations, this aim is served by theorems giving a classification of the linear groups having given properties. The choice of these properties is not, of course, accidental; indeed, the usual situation is that the appearance of theorems of this sort must satisfy some very concrete requirements.

One can give some general advice on how to attack the recognition problem. It is convenient to suppose that the ground field is algebraically closed; on the one hand, this is no burden since matrices over P can be viewed as matrices over its algebraic closure $\bar{P}$. On the other hand, transition from P to $\bar{P}$ can lead to some loss if G is closely connected with P itself. Nevertheless, we shall assume that $P = \bar{P}$.

The first step in studying a group G is to decide whether it is algebraic. In the majority of the important cases, the likelihood is that it will be. If not, then obviously one can count on using topological methods (§ 3), or bend efforts to the recognition of the Zariski closure $\overline{G}$ of G. Note that $\overline{G}$ is an algebraic group in the sense of 1.6.1.

So, we begin by supposing that G is an algebraic group (possibly finite). If G is reducible, its structure is determined largely by those of its irreducible components. If G is imprimitive, then basically everything reduces to a determination of imprimitivity systems and a study of the matrix subgroups fixing them. It is not always easy to carry out these two steps in practice, but when it can be done the problem is usually very much easier.

We give further consideration to the case of infinite G. The connected component G^0 is then nontrivial, and it usually has a decisive influence on the structure of G. However, if G^0 is abelian, it is central when G is primitive; in that case the structure of G is determined by a finite subgroup consisting of matrices of determinant 1. Suppose that G^0 is nonabelian. In this situation the structure of G is determined by its simple normal subgroups (see 1.6.2). Essentially, in view of the structure theorems from 1.6.2 and Clifford's theorem, everything reduces to the case where G^0 is simple. Since the simple algebraic groups have been classified (see 1.6.3), what we really want to do is to decide which of the irreducible representations is given in terms of highest weights (1.6.10). On the other hand, a fairly large number of diverse properties of algebraic group representations can be determined from the highest weights, especially in the characteristic 0 case. If properties known in advance can be compared with properties expressed in terms of highest weights, one can then hope for a complete identification of G^0. For example, if the matrices are of dimension 7, then G^0 is similar to $\mathrm{SL}_7(\mathbb{C})$ or $\mathrm{SO}_7(\mathbb{C})$ in their natural realisations, or else it gives a representation of $\mathrm{SL}_2(\mathbb{C})$ or $G_2(\mathbb{C})$. If G^0 contains a transvection, it is similar to $\mathrm{SL}_n(\mathbb{C})$ or $\mathrm{Sp}_{2n}(\mathbb{C})$ in the natural representations. If G^0 contains a matrix x such that $(x - E)^2 = 0$, we are dealing with a microweight representation (see [Bourbaki (1975), Chapter 8, 7.3]). Here is a more complicated example. If our group contains a matrix with all eigenvalues of multiplicity 1, it must be a realisation of a representation with highest weight as given by the following table (the notation is consistent with that in Bourbaki (1975):

$$A_n\colon m\overline{\omega}_1, m\overline{\omega}_n, \overline{\omega}_i \ (m = 1, 2, \ldots, i = 2, \ldots, n-1);$$

$$B_n\colon \overline{\omega}_1, \overline{\omega}_n; \qquad C_n\colon \overline{\omega}_1, \overline{\omega}_3 \quad \text{when} \quad n = 3; \qquad D_n\colon \overline{\omega}_1, \overline{\omega}_{n-1}, \overline{\omega}_n;$$

$$E_6\colon \overline{\omega}_1, \overline{\omega}_6, \qquad E_7\colon \overline{\omega}_1, \overline{\omega}_7; \qquad G_2\colon \overline{\omega}_1.$$

There is a description of the irreducible connected algebraic groups of prime characteristic containing matrices with simple eigenvalues (Zalesskij and I.D. Suprunenko 1987); the crucial case of tensor-indecomposable groups was considered earlier by Seitz (1987).

Consider now the case of finite G. Just as we did above, we attempt to reduce things to the case where G is irreducible and primitive. One can hope that the analysis of a primitive group can be reduced to that of a quasisimple subnormal subgroup in its irreducible representation, as described at the end of 2.3. Let F be a quasisimple irreducible group; set $S = F/Z(F)$, so that S is simple. Note that F differs from its commutator subgroup F' by a scalar only, so that it is enough to discover what sort of a group F' is. By a theorem of Schur (see [Curtis and Reiner (1962), Chapter VII, § 51]), F' is a representation of the *covering group*[1] $\tilde{S}$, which is determined by S and is such that $\tilde{S}/Z(\tilde{S}) = S$ and $\tilde{S}' = \tilde{S}$. The covering group $\tilde{S}$ is known for every known simple group S. If we proceed on the basis of the classification theorem for finite simple groups, which is apparently close to completion (so that we may assume that all simple groups are known), we then have to identify F' with an irreducible representation of the cover $\tilde{S}$ of some simple group S. The known simple groups consist of 26 sporadic groups, the alternating groups, and the finite simple Chevalley groups. There are character tables of the projective representations of the sporadic groups (see Conway, Curtis, Norton, Parker and Wilson (1972), for example). Thus, the identification of these groups can be achieved by calculating the traces of the matrices comprising the original group. The effort of doing this could prove laborious, but it will produce a result. There is an extensive representation theory of the Chevalley groups, and it is still being developed. Probably, the near future will see the discovery of formulae for the characters of these groups. In the meantime, this has been achieved already for the series $\mathrm{SL}_n(q)$, $G_2(q)$, ${}^2G_2(q^3)$, ${}^3D_4(q^3)$, $\mathrm{Sp}_{2n}(q)$ (here $n = 2$ or 3 for even q, $n = 2$ for odd q), and also for the non-simple groups $\mathrm{GL}_n(q)$ and $U_n(q)$. The possibility of identification emerges as a consequence of this work, at least in principle. The representation theory of the alternating group A_n is fairly well-developed, as part of that for the symmetric group. Much is known about the representations of $\tilde{A}_n$; the foundations were laid by Schur (1911).

We mention some other ideas and results that may be of use in this context. It seems that the case where the group is characterised by a small number of easily-verified properties is of particular value. We shall provide some examples where it is possible to identify a finite irreducible linear group from the knowledge of just one of its matrices. Recall that a *pseudoreflection* is a matrix of the form $E + t$, where t is a matrix of rank 1 and $n > 1$. It is easy to see that the Jordan form of a pseudoreflection is a matrix like $\mathrm{diag}(\varepsilon, 1, \ldots, 1)$ for $\varepsilon \neq 1$, or $\mathrm{diag}\left(\begin{pmatrix} 1 & 1 \\ 0 & 1 \end{pmatrix}, 1, \ldots, 1\right)$ (in the latter case we have a *transvection*). If $\varepsilon = -1$, we can omit the qualifier "pseudo" – this is often done in general as well, for simplicity. The following remarkable theorem is due in essence to Mitchell (1914).

Theorem A. *Let G be a finite irreducible primitive linear group over a field of characteristic 0, generated by psedoreflections. Up to similarity, G is isomorphic either to the symmetric group S_m in its representation of degree $m - 1$, or to one of 15 explicitly described matrix groups of degree at most 8.*

[1] Such groups are also called representation groups.

Observe that if the G in Theorem A is reducible, it is a direct sum (see (2.4)) of irreducible groups generated by pseudoreflections. There is an analogue of Theorem A for fields of prime characteristic (Wagner, Zalesskij, Serezhkin: see Zalesskij (1981)). With just one exception, the list consists of groups that are almost classical and groups obtained from those itemised in Theorem A on reduction to prime characteristic.

Let H be a finite primitive group, $H \subset \mathrm{GL}_n(P)$. If H contains a pseudoreflection, then H contains as irreducible normal subgroup a group G generated by pseudoreflections. The group G is known and its normaliser in GL_n is not very different from G. Therefore, H is practically known as well.

We denote the degree of the minimal polynomial of a matrix t by $\deg t$. If t is a pseudoreflection, then $\deg t = 2$. If t is diagonalisable, then $\deg t$ is the number of different eigenvalues. Quite a lot is known about finite linear groups containing matrices t with $\deg t = 2$. The characteristic 0 case will be considered in detail in 3.9.4 below. A description of the finite primitive linear groups of characteristic $p > 3$ containing elements t such that $(t - E)^2 = 0$ has been obtained by Thompson (1971). It turned out that such groups are the images of some representations of the Chevalley groups of characteristic p other than of type E_8. A list of these representations, as well as a good deal of other information that is sometimes helpful for identifying groups that contain such elements t, can be found in a paper of Premet and I.D. Suprunenko (1983). Thompson's work was extended to the case $p = 3$ in a series of papers by Ho in the years 1973–1976. He obtained lists of the groups arising, which are more diverse than Thompson's groups with $p > 3$, under certain restrictions which make his results less useful for recognising linear groups than is that of Thompson. Ho does not give explicitly the representations in which some element satisfies the equation $(t - E)^2 = 0$. Probably, in the majority of cases such a representation is one of smallest degree. We note that neither Thompson nor Ho used the classification of finite simple groups. There is no doubt that Thompson's result can be used to get a description of the primitive groups of prime characteristic $p > 3$ generated by elements t with $\deg t = 2$, $|t| > 3$ that are not necessarily unipotent. However, nobody has done this as yet.

Together with elements t such that $\deg t = 2$, another natural generalisation of pseudoreflections are the elements of the form $E + x$, where x is a matrix of rank 2. We shall call such elements two-dimensional pseudoreflections. Huffman and Wales obtained a description of the irreducible primitive linear groups over $\mathbb{C}$ generated by two-dimensional pseudoreflections, in a series of papers published between 1975 and 1980. These results have applications in invariant theory: every finite group G in $\mathrm{GL}_n(\mathbb{C})$ whose algebra of invariants is a complete intersection (more exactly, the algebraic variety is such) can be generated by two-dimensional pseudoreflections. Korlyukov has studied the analogous problem in positive characteristic (1983–1985).

Chapter 3
A Sketch of the Contemporary State of the Theory of Linear Groups

§ 1. Linear Groups from an Abstract Point of View

In recent decades, an important part of linear group theory has been that devoted to investigations carried out under the influence of the ideology of general group theory. How are linear groups to be distinguished amongst the abstract groups? Can linear groups be characterised in abstract terms: in other words, which groups have faithful linear representations? What about the structure of linear groups contained in some class of groups that is important in the abstract theory? The answer to the first two of these, at least, cannot be an easy one. From all angles, the linear groups constitute a very complex class.

1.1. "Arbitrary" Linear Groups. The main results on linear groups of the widest general nature were mentioned in §§4, 5 of Chapter 2, when we expounded the algebraic group method and the residual method. Up to the present day, only the groups of degree 2 have been studied in great detail.

Bass (1980) established the following theorem on the basis of results of Serre discussed at the end of 2.3.4.

Theorem. *Let P be a field that is finitely generated over its prime subfield, and $\bar{P}$ its algebraic closure. Let Γ be a subgroup of* $\mathrm{GL}_2(P)$ *and Γ_u the subgroup of Γ generated by the unipotent elements. One of the following conclusions holds*:

(a) *Γ/Γ_u has an infinite cyclic factorgroup*;

(b) *Γ is a free product $\Gamma_0 *_\Lambda \Gamma_1$ with amalgamated subgroup Λ such that $\Gamma_0 \neq \Lambda \neq \Gamma_1$ and every finitely generated unipotent subgroup of Γ is conjugate in Γ with a subgroup of Γ_0 or Γ_1*;

(c) *Γ is conjugate in* $\mathrm{GL}_2(\bar{P})$ *with a group of triangular matrices of the form $\begin{pmatrix} a & b \\ 0 & d \end{pmatrix}$, where a and d are roots of unity*;

(d) *Γ is conjugate with a subgroup of* $\mathrm{GL}_2(A)$, *where A is a subring of $\bar{P}$ whose additive group is finitely generated.*

In particular, if P has prime characteristic, A is a finite field, so that Γ is finite in case (d).

There are some general facts about the structure of soluble normal subgroups of linear groups. For example, the length of the upper central series of a linear group is always less than 2ω, where ω is the first infinite ordinal (see Wehrfritz (1973)).

1.2. Linearly Representable Groups. The first questions about linear representability of abstract groups were posed by Mal'tsev in 1940; he obtained the following result.

Theorem A. (a) *Let G be an abstract group and P any field. If every finitely generated subgroup of G can be embedded in* $\mathrm{GL}_n(P)$, *G itself can be so embedded.*

(b) *Let H be a finitely generated abstract group. Suppose that there exists, for each finite subset F of H, a normal subgroup N of H such that* $N \cap F = \emptyset$ *and* H/N *is embeddable in* $\mathrm{GL}_n(K)$ *for some field K depending on F. Then, H is embeddable in* $\mathrm{GL}_n(L)$ *for a suitable field L.*

The direct sum construction for linear groups (see 1.2.4) shows that the direct product of linearly representable groups has a linear representation. The proof of the next result is much harder than this.

Theorem B (Nisnevich 1940). *Let G be the free product of groups* G_μ $(\mu \in M)$. (a) *If each of the* G_μ *has a faithful representation in* $\mathrm{PGL}_n(P)$ *for fixed n and some field P, then G has a faithful representation in* $\mathrm{PGL}_n(P_1)$ *for some extension* P_1 *of P.* (b) *If each* G_μ *has a faithful representation in* $\mathrm{GL}_n(P)$, *then G has a faithful representation in* $\mathrm{GL}_{n+1}(P_1)$ *for some extension* P_1 *of P.*

Note that (b) follows from (a). As far as we know, the natural problem about extending (a) to an arbitrary simple algebraic group is still unsolved.

A free product with amalgamated subgroup need not have a linear representation even if the factors do. The discovery of criteria for such groups to have linear representations is a problem that has not yet been solved in full, though there are many partical results.

Mal'tsev also produced a criterion for abelian groups to be linear. The following definition will enable us to state it. We say that a group G has (Mal'tsev) *rank r* if every finite subset R of G is contained in a subgroup with r generators. We denote by $\pi(G)$ the maximal periodic normal subgroup of G.

Theorem (Mal'tsev 1940). *An abelian group A can be represented by* $n \times n$ *matrices over a field of characteristic* 0 *if and only if the rank of* $\pi(A)$ *does not exceed n. A condition for representability over a field of positive characteristic p is that the Sylow p-subgroup S of* $\pi(A)$ *have finite exponent* p^l *and* $r + l \leqslant n$, *where r is the rank of* $\pi(A)/S$.

Recently (in 1988), Lubotsky made a significant step forward in the understanding of the linear representability problem. We make the following definition in order to expound it. Let c be any natural number and p any prime. We say that an abstract group Γ has a p-congruence structure of weight c if Γ has a series $\Gamma = N_0 \supset N_1 \supset \cdots \supset N_i \supset \cdots$ of normal subgroups of finite index such that (i) $\bigcap_i N_i = \{1\}$; (ii) N_1/N_i is a finite p-group for $i = 2, 3, \ldots,$ and (iii) each N_j/N_i $(1 \leqslant j < i)$ can be generated by c elements.

Theorem D. *Let* Γ *be a finitely generated group. Then,* Γ *has a faithful linear representation over* $\mathbb{C}$ *if and only if* Γ *has a p-congruence structure of some weight c.*

There is a connection between the number c and the degree of the linear representation of N_1. The proof of the theorem uses the theory of p-adic Lie groups (see 1.4.5). The first step is this: using the characterisation of p-adic Lie groups in the class of pro-p-groups, one proves the following result:

A group Γ has a p-congruence structure of finite weight if and only if Γ can be embedded in a compact p-adic Lie group.

Next, it follows from the theory of p-adic Lie groups that every compact p-adic Lie group can be represented by matrices over the field of p-adic numbers. For the transition to $\mathbb{C}$, one notes that every finitely generated field of characteristic 0 is embeddable in $\mathbb{C}$.

Note that Lubotsky's theorem is ineffective in solving the problem of when a group can be represented by matrices over the rationals. We have the following result in this direction:

Theorem E (Swan 1967; Auslander 1967). *Every polycyclic group can be represented by integral matrices.*

Conditions under which finitely generated groups can be represented by matrices over fields of positive characteristic are yet to be discovered. Almost nothing is known about representability of infinite nonabelian p-groups by matrices over fields of characteristic p.

1.3. The Main Classes of Abstract Groups and Linear Groups. During the process of investigations on the general group concept, certain classes of groups appeared that went on to receive special attention and thorough discussion. We have in mind here the soluble, nilpotent and periodic groups; the p-groups, groups satisfying identical relations *etc.* What does the intersection of each of these classes with the class of linear groups look like? For instance, what are the abstract properties of soluble or periodic linear groups? The normal situation is that every class of this sort is quite easy to describe in abstract terms. We know already that a linear group satisfying an identical relation is soluble-by-finite (§ 2.4), and that every soluble linear group is nilpotent-by-abelian-by-finite (2.3.3). The nilpotency class does not exceed the degree of the matrices concerned, and the index is bounded by a quantity depending only on the degree. This extra information is rather less substantial from the abstract viewpoint, however. To continue: a linear p-group is abelian-by-finite if p is different from the characteristic of the ground field; if p is the same as the characteristic, ever linear p-group is nilpotent, and indeed is similar to a subgroup of the unitriangular group. A locally nilpotent linear group need not be nilpotent, but always has a nilpotent normal subgroup of finite index. When the ground field is of characteristic not zero, the commutator subgroup of a locally nilpotent linear group is periodic. Periodic linear groups are locally finite (see 2.2.4); for each p, their Sylow p-subgroups are conjugate. These facts, together with many others of this sort, can be found in the books of D.A. Suprunenko (1972) and Wehrfritz (1973).

Thus it is evident that imposition of linearity on a class of groups can bring about severe restrictions on it. Let us stay with periodic linear groups for the moment. With $\overline{\mathbb{Z}}_p$ denoting the algebraic closure of $\mathbb{Z}_p$, every linear group over $\overline{\mathbb{Z}}_p$ is periodic. Using the classification of finite simple groups, it can be shown that every infinite periodic simple linear group is a Chevalley group over some subfield of a $\overline{\mathbb{Z}}_p$. There is a proof of this result for fields of characteristic not 2 that does not depend on the classification theorem (Borovik 1984).

§ 2. Conjugacy Theorems for Linear Groups

If two subgroups H_1 and H_2 are conjugate in the general linear group $\mathrm{GL}(V)$, they are represented by the same matrices with respect to suitable choices of bases. Thus, from many points of view they are indistinguishable. What this remark means is that the H_1-module V and the H_2-module V are isomorphic. Therefore, it is understandable that there is a tendency to describe linear objects, in particular linear groups, up to conjugacy in the general linear group. Note that this theme has connections with the problem of classifying certain types of linear group, for example the abelian groups. In this section we shall touch on problems concerning conjugacy under certain subgroups of the general linear group.

2.1. Conjugacy Theorems for $\mathrm{GL}_n(P)$: the Case of an Algebraically Closed Field

(A) Every abelian subgroup is similar to a subgroup of the group of upper-triangular matrices.

This assertion is equivalent to the following well-known fact. If A is an abelian subgroup of $\mathrm{GL}(V)$, there is an A-module W of dimension 1 contained in V.

(B) Every abelian group consisting of diagonalisable matrices is itself diagonalisable, that is, is similar to a group of diagonal matrices.

(C) (Platonov 1966, Kolchin 1948). Every linear group satisfying an identical relation (in particular, every soluble linear group) has a triangularisable subgroup of finite index.

(D) (D.A. Suprunenko 1955). The maximal locally nilpotent subgroups of $\mathrm{GL}_n(P)$ fall into finitely many conjugacy classes.

(E) (D.A. Suprunenko 1962). If P is of characteristic zero, the maximal periodic subgroups of $\mathrm{GL}_n(P)$ fall into finitely many conjugacy classes.

(F) (Schur 1911). Every periodic subgroup (in particular, every finite subgroup) G of $\mathrm{GL}_n(\mathbb{C})$ is similar to a subgroup of the unitary group $U_n(\mathbb{C})$.

Some of these results are proved within the general context of algebraic groups, as in 1.6.4. We note further:

(G) (Platonov 1969). The maximal soluble subgroups of an algebraic group form finitely many conjugacy classes.

In many applications, it is desirable to have a description (up to conjugacy) of the abelian subgroups A of $\mathrm{GL}_n(P)$, or at least of those that are Zariski-closed, or Euclidean-closed when $P = \mathbb{C}$. Unfortunatey, it is probable that no effective solution to this problem exists. For example, the situation could be regarded as satisfactory if at least the maximal abelian subgroups of a linear group can be described. However, not even this much more specialised question is amenable to solution (see 3.2 below).

2.2. Conjugacy Theorems for $\mathbf{GL}_n(P)$: the Arbitrary Field Case. There are a number of results about the conjugacy of certain type of subgroups of $\mathrm{GL}_n(P)$ when P is quite arbitrary. The most important are associated with the theory of algebraic groups. For instance, the following results were obtained by Borel and Tits (1965).

Theorem. *Let P be a perfect field and G a connected algebraic linear group defined over P. Then, the maximal connected unipotent subgroups of G_P are conjugate in G_P. The parabolic subgroups of G_P fall into finitely many conjugacy classes. The minimal parabolic subgroups are conjugate.*

We recall that a parabolic subgroup of G is one that contains a Borel subgroup (see 1.6.1). A parabolic subgroup of G_P is a subgroup of the form $\Pi \cap G_P = \Pi_P$, where Π is a parabolic subgroup of G defined over P.

We record some more specialised results at this point.

(A) Set $p = \operatorname{char} P$, and let m be a natural number not divisible by p. The number of conjugacy classes in $\mathrm{GL}_n(P)$ of subgroups of order m is finite.

(B) The irreducible subgroups of fixed finite order fall into finitely many conjugacy classes in $\mathrm{GL}_n(P)$.

These results are proved in courses on the representation theory of finite groups. Note that the irreducibility condition in (B) is essential.

(C) (Vol'vachev 1963). For $p \neq 2$, the Sylow p-subgroups of $\mathrm{GL}_n(P)$ are conjugate. This is also true for $p = 2$ if P contains a square root of -1.

For arbitrary P, the Sylow 2-subgroups of $\mathrm{GL}_n(P)$ can be described compltely; despite this, they are not always conjugate.

Assertions (A) and (C) do not in general carry over to subgroups of $\mathrm{SL}_n(P)$. The following, typical, example is borrowed from Platonov (1966). Suppose that $P = \mathbb{Q}$ and set

$$g_i = \begin{pmatrix} 0 & -p_i \\ p_i^{-1} & 0 \end{pmatrix},$$

where the p_i are different primes. Then $\{g_i\}$ is a group of order 4, and straightforward verification shows that $\{g_i\}$ and $\{g_j\}$ are not conjugate in $\mathrm{SL}_2(\mathbb{Q})$ for $i \neq j$ (but they are conjugate in $\mathrm{GL}_2(\mathbb{Q})$). If the p_i run over the primes in the arithmetic progression $4k + 3$, $k = 1, 2, \ldots$ (Dirichlet's Theorem asserts that there are infinitely many such), then the $\{g_i\}$ are Sylow 2-subgroups of $SL_2(\mathbb{Q})$.

2.3. Conjugacy in Subgroups of $\mathrm{GL}_n(P)$. Some of the results mentioned above carry over to the classical groups. Generally speaking, it is difficult to see any pattern in questions about conjugacy of subgroups in linear groups of any general nature. Nevertheless, there are some results of significant generality. One of these is due to Borel and Harish-Chandra (1962), and is a consequence of their reduction theory for arithmetic groups.

Theorem. *Let G be an algebraic group defined over an algebraic number field P, and H any arithmetic subgroup of G. There are only finitely many conjugacy classes of finite subgroups in H.*

This result was extended by Borel and Serre (1976) to the so-called S-arithmetic subgroups, where S is a finite set of non-Archimedean valuations on P. In particular, it follows from the theorem of Borel and Harish-Chandra that there are only finitely many conjugacy classes of finite subgroups in $\mathrm{GL}_n(\mathbb{Z})$ and $\mathrm{Sp}_{2n}(\mathbb{Z})$; the Borel-Serre theorem gives the same result for $\mathrm{GL}_n(\mathbb{Z}[1/m])$ and $\mathrm{Sp}_{2n}(\mathbb{Z}[1/m])$, where m is any natural number.

In 2.3.3. we gave a proof of Platonov's theorem on the conjugacy of the Sylow p-subgroups of periodic linear groups. Platonov (1966) also established that the maximal π-subgroups of soluble periodic linear groups are conjugate, where π is any set of prime numbers. By definition, a *π-group* is a group whose elements are of finite order divisible by primes in π only. These results are generalisations of the classical theorems of Sylow and P. Hall for finite groups (see also Wehrfritz, 1968). One could ask in which linear groups the classes of conjugate elements have a complete description. If we forget certain very special cases, this has been done for the classical groups over an arbitrary field (Wall 1963), for the finite Chevalley groups, and also for semisimple algebraic groups over algebraically closed fields.

§ 3. Disposition of Subgroups

In this section we consider questions about maximal subgroups, chains of subgroups *etc.* To some extent, they clarify our understanding of the structure of the lattice of subgroups; however, it would of course be unrealistic to expect exhaustive results.

3.1. Maximal Subgroups. Currently, there is a fairly rich collection of examples of maximal subgroups of the general linear group and other classical groups. However, it is unclear how far these examples are typical of the class of maximal subgroups, and it is even less clear how to find a description of all of them.

It is convenient to expound this topic in the more general context of algebraic linear groups. The following theorem of Borel and Tits (1965) is among the most general, since it is valid for nearly all algebraic groups over arbitrary fields.

Theorem A. *Let G be a connected algebraic linear group defined over a field P. The maximal parabolic subgroups of G_P are maximal in G_P as abstract groups.*

Note that this theorem is significant only when G is insoluble (otherwise, G has only one parabolic subgroup, namely G itself). As an example, in the orthogonal, symplectic or unitary group, every maximal parabolic subgroup is the stability group of some isotropic subspace, and conversely. We recall at this point a "non-maximality" theorem, which plays an important part in finite group theory when the field P is finite:

Theorem B (Borel and Tits 1971). *Let G be a reductive group defined over a field P, and U a unipotent subgroup of G_P. There exists a parabolic subgroup Π of G defined over P such that*

(a) *the normaliser of U in G_P is contained in Π_P,*

(b) *U is contained in a unipotent normal subgroup of Π_p.*

There is a fairly general result relating to the maximality of the Chevalley group G_K in G_P, where K is a maximal subring of P that is not a field.[2] The point is that the Iwasawa and Cartan decompositions hold for such subrings.

Proposition. *Let B be a Borel subgroup and H a Cartan subgroup of a split (= normal type) Chevalley group G_P. Then $G_P = G_K \cdot B$ (the Iwasawa decomposition) and $G_P = G_K H G_K$ (the Cartan decomposition).*

Let us explain. Steinberg (1967, § 8) shows that decompositions of these sorts exist in the case where K is a principal ideal ring. In reality, his argument can be improved so as to work for any valuation ring. The next result is easy to deduce from this.

Theorem C. *Let K be a maximal subring of a field P that is not a field. Then, the Chevalley group G_K is maximal in G_P.*

Suppose that $g \in G_P$, $g \notin G_K$, and that the requirements of the theorem hold. By the Cartan decomposition, we can assume that $g \in H$. Thus, as is easy to see, $gx_\alpha(K)g^{-1}$ is not contained in $x_\alpha(K)$ for any root subgroup $x_\alpha(K)$. This reduces everything to groups of type A_1. In the case of $\mathrm{SL}_2(P)$, the situation is easy to analyse from first principles. Note that Theorem C was proved for $\mathrm{SL}_n(P)$ by Schmidt in 1981.

Essentially, Theorems A and C are of a common nature. The fact is that G_K can be interpreted as a maximal parabolic subgroup of G_P with respect to a Tits system (G_P, B_I, N), where N is the normaliser of H in G_P, B_I is generated by B and the groups $x_\alpha(I)$, $\alpha < 0$, and I is a maximal ideal of K (see [Borel and Tits (1965), Chapter IV, § 2, Definition 1 and Theorem 3; see also exercise 21]).

Theorem C yields the following not uninteresting qualitative fact.

---

[2] In § 6.8 of Chapter 1, we restricted the definition to finite Chevalley groups. The reader can take this definition for infinite fields as meaning that a decomposable group has a structure similar to that of a simple algebraic group over an algebraically closed field. However, familiarity with the theory of the Chevalley groups is very desirable. We have in mind the canonical (root) embedding of G_K in G_P; see the explanation preceding Lemma 48 in Steinberg (1967).

Corollary. *The group $G_{\mathbb{R}}$ contains subgroups that are not Euclidean-closed but are maximal as abstract subgroups. If P is not a union of finite fields, then G_P contains subgroups that are not Zariski closed but are maximal as abstract subgroups. Moreover, every finitely generated subgroup of G_P is contained in infinitely many maximal subgroups that are not Zariski closed.*

To explain: the groups G_K, where K is a maximal subring of P that is not a field are subgroups of this type. Subrings of that sort can be constructed as valuation rings of rank 1 of P (see [Bourbaki (1961–1965), Chapter IV, §4, subsection 5]) that are suitable extensions of non-Archimedean valuation of the field $\mathbb{Q} \subset P$ (if char $P = 0$) or of $\mathbb{Z}_p(t)$ if char $P = p > 0$, where now t is an element of P transcendental over $\mathbb{Z}_p$.

When K is a maximal subfield of P and G_K is canonically embedded inG_P, G_K is also close to being maximal. Usually, the normaliser of G_K in G_P is maximal. Exceptions occur only for an imperfect field of characteristic 2, or if G is of type G_2 of characteristic 3; the construction of the intermediate groups was indicated by Steinberg (see [Steinberg (1967), the remark preceding Theorem 29 in §10]). The case of a finite field was investigated in detail by Burgoyne, Griess and Lyons (1977). For infinite P, the question was considered by Bashkirov (1984) in relation to $\mathrm{SL}_2(P)$, and by Nuzhin (1983) for groups other than $\mathrm{SL}_2(P)$.

We note the following simple fact:

Proposition. *Let P be an algebraically closed field, and G, H algebraic groups over P such that $G \subset H$, H is simple, and G is infinite and maximal in H as algebraic subgroup. In these circumstances, G is maximal in H as abstract subgroup.*

Proof. Let G^0 be the connected component of G. Suppose that F is an intermediate group, that is, $G \subset F \subset H$ and $F \neq H$. By Theorem E in 2.3.3, the group F_1 generated by all the groups fG^0f^{-1} as f runs over F, is closed and connected in the Zariski topology. Furthermore, $G^0 \subset F_1 \subset F_1G \subset F$ and F_1 is of finite index in F_1G. Thus, F_1G is a closed subgroup, so that $F_1G = G$ and $F_1 = G^0$ since G is maximal. Hence G^0 is normal in F, and therefore in the closure E of F, which is just H since G is maximal. This contradicts the simplicity of H provided that $H \neq G$ (because proper closed normal subgroups of a simple algebraic group are finite).

The natural question as to whether this proposition can be extended to the groups of K-points of algebraic groups, K an infinite subfield of P, usually has a negative answer. However, all examples known to the author relate to the case where H_K is not simple as abstract group, apart from certain anomalies for imperfect fields of characteristic 2 or 3 mentioned by Steinberg (1967, §10). It is possible that abstract simplicity of H_K is a reasonable restriction for all other fields; under that assumption, it can be expected that the normaliser of G_K in H_K is a maximal subgroup of H_K. Certain results in this direction are covered in the author's survey[3] (Zalesskij (1983), §7). As a typical special case, we have the following question:

Let ϕ_n: $\mathrm{SL}_2(\mathbb{Q}) \to \mathrm{GL}_n(\mathbb{Q})$ be the natural representation of $\mathrm{SL}_2(\mathbb{Q})$ in the space of polynomials of degree n (see 1.2.14). Set $G_n = \phi_n(\mathrm{SL}_2(\mathbb{Q}))$. If n is even, G_n is contained in a unique symplectic group $\mathrm{Sp}(f, \mathbb{Q})$. Let N be the normalizer of G_n in $\mathrm{Sp}(f, \mathbb{Q})$. Is N a maximal subgroup of $\mathrm{Sp}(f, \mathbb{Q})$?

Margulis and Soifer developed the Tits techniques and proved the following interesting result (1981):

Theorem D. *If all the maximal subgroups of a finitely generated linear group are of finite index, then G has a soluble subgroup of finite index.*

Here is the idea of the proof. Use the Tits method to construct a proper free subgroup S of G having nonempty intersection with every set of the form gH, where $g \in G$ and H is a subgroup of finite index in G. If M is a maximal proper subgroup containing S, then because of the construction of S, M cannot be of finite index in G. Moreover, the cardinal of the set of maximal subgroups of infinite index is uncountable.

A description of the maximal connected subgroups of simple algebraic groups of characteristic 0 follows from the papers of Dynkin (1952) on the maximal subalgebras of simple Lie algebras. The analogous problem for fields of prime characteristic was solved by Seitz (1987) and Testerman (1988). The results are employed in the determination of the maximal subgroups of the finite Chevalley groups (Liebeck, Saxl and Seitz, 1987).

3.2. Relatively Maximal Subgroups. The object here is to discuss the linear groups that are maximal among the groups contained in some special class or other. For instance, the maximal abelian, maximal soluble and maximal periodic linear groups, *etc*, have been investigated. The consideration of the maximal subgroups is one of the important approaches to the determination of the structure of the groups in a given class, since there are usually too many individual groups to study them all separately. At the same time, maximal subgroups can in fact carry quite a lot of information about the totality of groups under consideration.

(A) Maximal abelian groups. Let A be a maximal abelian subgroup of $\mathrm{GL}_n(P)$, and $[A]_P$ its enveloping algebra. Clearly, A is the group of invertible elements in $[A]$. Moreover, in almost all cases (exceptions arise only when $|P| = 2$), $[A]$ is a maximal commutative subalgebra of $M_n(P)$. Therefore, the problem of determining the maximal abelian groups comes down to that of finding the maximal commutative subalgebras of the full matrix algebra; indeed, it is practically equivalent to it. The latter question reduces without difficulty to finding the maximal commutative algebras of upper-triangular matrices with zero main diagonal.

The central result here is the classical theorem of Schur (1905) stating that the

[3] In the statement of the unsolved maximality problem in [Zalesskij (1983), p. 152], one has to assume in addition that H_K is generated by unipotent elements. There, instead of $\mathrm{GL}(2n, k)$ and $\mathrm{GL}(n, k)$ one should read $\mathrm{SL}(2n, k)$ and $\mathrm{SL}(n, k)$.

dimension of a commutative algebra of matrices of degree n does not exceed the integer part of $n^2/4$; a commutative algebra of that dimension exists for each n. Currently, a complete description of the maximal commutative subalgebras seems to be out of sight, and investigators have decided that the best course is to sort out special classes of algebras. The reader will find more details in D.A. Suprunenko and Tyshkevich (1966). Observe that for $n \leqslant 6$ there are only finitely many similarity classes of maximal commutative algebras of matrices of degree n over an infinite field, while there are infinitely many for $n > 6$ (D.A. Suprunenko 1956).

(B) Maximal soluble linear groups. The principal worker here is D.A. Suprunenko, and the main results are expounded in D.A. Suprunenko (1958). This is how the general picture looks. Firstly, the problem reduces to finding the maximal soluble primitive linear groups, and to finding the maximal soluble subgroups of the symmetric groups. This last question comes down to describing the maximal soluble linear groups over finite fields. The problem of describing the primitive groups reduces completely to that for the maximal soluble subgroups of the symplectic group over a prime field. It seems that one cannot reduce this problem to anything simpler; besides, soluble subgroups of the symplectic group can be embedded in maximal soluble subgroups of the general linear group over a finite field, and the reduction continues. In this way one can deduce various estimates for the arithmetic invariants of soluble linear groups, things like the solubility length.

(C) Maximal locally nilpotent linear groups. This class of groups has been described in detail by D.A. Suprunenko: see [D.A. Suprunenko (1972) Chapter VII]. It is enough to consider the indecomposable case. It turns out that every indecomposable maximal locally nilpotent linear group over a perfect field splits as a Kronecker product of an upper-triangular group (more exactly, a conjugate of such a group) and an irreducible maximal locally nilpotent group. The structure of groups of this last type is well known (D.A. Suprunenko (1972)). For example, they split as Kronecker products of groups whose central factor-groups are p-groups for certain primes p dividing the matrix degrees. D.A. Suprunenko (1972) also contains a description of the maximal irreducible nilpotent linear groups of given nilpotency class. The fundamental fact here is that the commutator subgroups and central factor-groups of such groups are finite. There are estimates in D.A. Suprunenko (1972) for the order of the central factor-group in terms of the nilpotency class and the degree of the matrices.

(D) The Sylow p-subgroups of $\mathrm{GL}_n(P)$. The main result on this topic is due to Vol'vachev (1963). The determination of the reducible Sylow p-subgroups comes down to that of the irreducible ones, and thereafter to the primitives. More exactly, every irreducible group can be written as a matrix wreath product (see 1.2.6) of a primitive Sylow p-subgroup of the general linear group of a lower degree and the Sylow p-subgroup of some symmetric group. For $p > 2$, the primitive linear p-groups are cyclic, while for $p = 2$ they have cyclic subgroups of index 2. Moreover, for $p > 2$ all the Sylow p-subgroups of $\mathrm{GL}_n(P)$ are conjugate: this for arbitrary P.

(E) Maximal periodic subgroups of $\mathrm{GL}_n(P)$. For char $P \neq 0$, there is a complete description of these subgroups (Zalesskij 1966). Let G be a subgroup of $\mathrm{GL}_n(P)$ of this type; if G is irreducible, then G is similar to $\mathrm{GL}_n(P_a)$, where P_a is the union of the finite subfields of P. If $P \neq P_a$, the conjugacy classes of maximal periodic subgroups are in correspondence with the conjugacy classes of parabolic subgroups, and there are 2^{n-1} of them.

If char $P = 0$, the situation is quite different. We note first that description of the subgroups under consideration reduces easily to that of the irreducible and the primitive cases. However, further progress meets with real difficulties. Let us consider the case $P = \mathbb{C}$ as an example. In this case a primitive periodic group has a scalar subgroup of finite index, and the commutator subgroup is finite. In essence, the problem comes down to describing the maximal finite subgroups of $\mathrm{SL}_n(\mathbb{C})$. This has been done for $n < 10$: see 3.4.5 below. In general, however, the problem has been studied but little. Examples of maximal finite subgroups of $\mathrm{SL}_n(\mathbb{C})$ can be extracted from various characterisation theorems for finite linear groups.

3.3. Intermediate Subgroups. There are a number of results about the determination of the subgroups of $\mathrm{GL}_n(P)$ and other classical groups containing a given subgroup. Occasionally, additional requirements are imposed so as to restrict the problem and obtain more definitive answers.

For instance, with $|P| > 2$, there is a description of the subgroups of $\mathrm{GL}_n(P)$ containing the group $D_n(P)$ of diagonal matrices. Let G be a subgroup such that $D_n(P) \subset G \subset \mathrm{GL}_n(P)$. It is easy to see that if G is reducible, it is conjugate *via* a permutation matrix (see 1.1.6) to a block triangular matrix group with irreducible diagonal blocks G_{ii}. It turns out that the off-diagonal blocks G_{ij} are either zero or complete sets of matrices of the requisite dimension, and also that $G_{ik} \neq \{0\}$ if $G_{ij} \neq \{0\} \neq G_{jk}$. In this way, the determination of intermediate subgroups is reduced to the case of irreducible groups. A group of the latter type can be written as a matrix wreath product (see 1.2.6) of a primitive group containing the diagonal by a transitive subgroup of the symmetric group of suitable degree (clearly, the transitive subgroup can be quite arbitrary). The primitive groups $H \subset \mathrm{GL}_m(P)$ containing $D_m(P)$ look like this. If $|P| > 5$, then $H = \mathrm{GL}_m(P)$; when $|P| = 5$, there is an exception when $m = 2$ and $|H| = 2^5 \cdot 3$; for $|P| = 4$ and 3 the list is longer (Borevich and Koibaev, 1976–1981).

For $|P| > 5$, these results extend to classical groups of maximal index (Vavilov, 1980–1982). It must be noted that the structure of the diagonal subgroups in the orthogonal and unitary groups depends crucially on the choice of basis. Vavilov's papers treat only groups of maximal index (namely $[n/2]$, where n is the matrix degree), and the basis chosen is a Witt basis (see 1.3.5). The contrary case – where the basis is orthogonal – has been considered for the unitary group over a local field by Krupetskij (1979). Seitz has investigated the problem of determining the subgroups of a finite Chevalley group containing a maximal torus, in a series of articles in the period 1979–1983. In this direction, subgroups that contain a maximal (nonsplit) torus properly are expected to be rare exceptions. However, Koibaev (1990) has observed that this is not so in general. His examples are concerned with subgroups of $\mathrm{GL}_2(\mathbb{Q})$ containing the torus $T = \begin{bmatrix} a & bm \\ b & a \end{bmatrix}$ with $a, b, m \in \mathbb{Q}$, where m is fixed and $m^{1/2} \notin \mathbb{Q}$. More precisely, let $G(\sigma)$ be the net group (see 1.1.8) where $\sigma_{21} = \sigma_{12}m \supseteq \sigma_{11} = \sigma_{22} \supseteq \sigma_{12}$. Then T normalizes $G(\sigma)$ and $\langle T, G(\sigma)\rangle$ is a subgroup of $\mathrm{GL}_2(\mathbb{Q})$.

It should be noted that the problem of describing the subgroups of classical groups containing the diagonal subgroups is usually studied in a more general context, namely when P is a (usually semilocal) ring. Details can be found in the survey Zalesskij (1983).

In fact, some of the results mentioned in 3.1 were proved in somewhat greater generality. Thus, if G_P is a Chevalley group over a field P, there is a description of the subgroups H intermediate between G_K and G_P, where K is a subfield of P such that the extension P/K is finite (or algebraic). The problem of finding intermediate subgroups has also been considered in the case where P is the field of fractions of the subring K.

Other results known to us are some relating to finite linear groups. The investigation of the maximal subgroups of finite groups, in particular of simple groups, is a traditional theme in finite group theory. The specific character of linear groups shows up in such problems as the determination of the inclusions between irreducible groups. Important contribution is Seitz's paper (1990) where he determines all pairs $X \subset Y$ of simple Chevalley groups of the same characteristic p such that for some representation ϕ of Y over a field of characteristic other than p the group $\phi(X)$ is irreducible. We mention the following interesting example. Let $n = n_1, n_2, \ldots, n_r$ be different natural numbers such that $n_{i+1}/n_i \in \mathbb{N}$ for $i = 1, \ldots, r-1$. Let q be an odd prime. Then, in every irreducible complex representation ϕ of degree $(q^n \mp 1)/2$ of $\mathrm{Sp}_{2n}(q)$, all the subgroups

$$\mathrm{Sp}_{2n}(q) \supset \mathrm{Sp}_{2n_2}(q^{n_1/n_2}) \supset \mathrm{Sp}_{2n_3}(q^{n_1/n_3}) \supset \cdots \supset \mathrm{Sp}_2(q^n)$$

are irreducible. Therefore, there exist arbitrarily long series of irreducible inclusions of Chevalley groups in $\mathrm{GL}_m(\mathbb{C})$, for suitable m. However, it is possible that this example is the only one of its type.

The question of the minimal irreducible groups is a natural one; these are the irreducible groups in which all proper subgroups are reducible. As Platonov has shown (1975), such groups are always finite. The minimal soluble groups in $\mathrm{GL}_n(P)$, where n is a prime or the product of two primes, were considered by D.A. Suprunenko in 1972.

§4. Finite Linear Groups

There is a substantial difference between the structure and properties of finite linear groups over fields of characteristic zero and fields of prime characteristic. Striking evidence of this is Jordan's theorem (see 2.3.1), which gives in particular that every subgroup G of $\mathrm{GL}_n(\mathbb{C})$ whose order is sufficiently large in comparison with n must have a nontrivial abelian normal subgroup. Nothing like this can be expected over fields of prime characteristic, as is shown by the following obvious example. Let P be an algebraically closed field of characteristic $p > 0$. Then, for every finite field F of characteristic p, we have $\mathrm{GL}_n(F) \subset \mathrm{GL}_n(P)$, and all the abelian normal subgroups of $\mathrm{GL}_n(F)$ are scalar for $n > 1$ with $n|F| > 4$. The order of $\mathrm{GL}_n(F)$ is $(q^n - q^{n-1})(q^n - q^{n-2})\dots(q^n - 1)$, and it increases without limit as the order q of F increases. This example makes one think that the overall behaviour of finite linear groups over fields of zero and positive characteristic must be very different from each other. We shall find plenty of corroboration for this thought in what follows. The reader should be particularly alert for references to the characteristic of the ground field in this section.

4.1. Invariant Hermitian and Bilinear Forms. Let V_n be a vector space of dimension n, with $\mathrm{GL}_n(\mathbb{C})$ acting on it in the natural fashion. As we saw in 1.3.13, for every subgroup G of $\mathrm{GL}_n(\mathbb{C})$ there is a nondegenerate positive definite Hermitian form on V_n invariant under G. This result is obtained by a typical device often called the averaging method. Let f be any nondegenerate positive definite Hermitian form on V_n and set $f_G = \sum_{g \in G} gf$, where $gf(x, y) = f(gx, gy)$ for x, y in V_n. Clearly, the form f_G is nondegenerate, Hermitian, positive definite, and invariant under the action of G. It is also clear that this fact is equivalent to the statement that G is similar to a subgroup of $U_n(\mathbb{C})$ (see 1.3.13). If G is a subgroup of $\mathrm{GL}_n(\mathbb{R})$, the averaging method yields the existence of a bilinear symmetric positive definite form invariant under G.

The following question arises here. Is there always a bilinear nondegenerate G-invariant form on V_n? The answer is no, even for $n = 1$ if $|G| > 2$. If M is the matrix of the bilinear form, G-invariance means precisely that ${}^t g^{-1} = MgM^{-1}$ for all g in G (see 1.3.6). Therefore, a necessary condition for the existence of an invariant bilinear form is that every matrix g in G is similar to ${}^t g^{-1}$. For the complex field, and any field at all if G is absolutely irreducible, the similarity condition is also sufficient, since the representations $g \to g$ and $g \to {}^t g^{-1}$ are themselves equivalent, because they have the same character (1.2.8); that is, $MgM^{-1} = {}^t g^{-1}$ for all g in G and suitable M in $\mathrm{GL}_n(P)$. It is clear that this M can be taken as the matrix of a G-invariant form.

Let D be the centraliser of G in the matrix ring $M_n(\mathbb{C})$. If M represents the matrix of an invariant form, then dM and Md are also matrices of G-invariant forms, for each d in D. In other words, the space N of G-invariant forms is a D-module. If M and M_1 are nondegenerate matrices in N, then $MgM^{-1} = M_1 g M_1^{-1}$, whence it follows that $M_1^{-1}M \in D$. This means that the nondegenerate G-invariant forms constitute an orbit of the group $D^* = D \cap \mathrm{GL}_n(\mathbb{C})$. If G is irreducible, then D consists of scalar matrices (Schur's Lemma, 2.1.3), that is, in this case the nondegenerate G-invariant bilinear forms differ only by factors from $\mathbb{C}$. We write $M = M^+ + M^-$, where M^+ is symmetric and M^- is skewsymmetric. This decomposition is unique, so that M^+ and M^- are G-invariant if M is. If G is irreducible, then M is M^+ or M^-. In other words, if f is a bilinear form invariant under an irreducible subgroup G of $\mathrm{GL}_n(\mathbb{C})$ (not necessarily finite), then f is symmetric or skewsymmetric.

There is a beautiful criterion due to Frobenius for the existence of an invariant bilinear form for a finite irreducible subgroup G of $\mathrm{GL}_n(\mathbb{C})$. Set $\Phi_G = \sum_{g \in G} \mathrm{tr}(g^2)/|G|$.

Theorem (Frobenius). $\Phi_G \in \{0, 1, -1\}$. *Moreover, G has a nondegenerate invariant bilinear form f if and only if $\Phi_G \neq 0$; f is symmetric if $\Phi_G = 1$ and skewsymmetric if $\Phi_G = -1$.*

In practice, the construction of the G-invariant form can present great difficulty. A device which sometimes helps is to go over to some – possibly smaller – subgroup G_1 of G with the same centraliser D; as indicated above, the set of nondegenerate invariant forms is undiminished, while the volume of work does reduce, thanks to the reduction in the order of the group.

For a finite linear group G over a field P of positive characteristic p, a condition for the existence of an invariant bilinear form is that the representations $g \to {}^t g^{-1}$ and $g \to g$ be equivalent. There may not be a Hermitian form.

4.2. Reducibility of Finite Linear Groups. The fundamental qualitative result here is Maschke's Theorem (more exactly, the Molien-Maschke Theorem) on the complete reducibility of finite linear groups over fields of characteristic 0 (the theorem follows immediately from the existence of an invariant positive definite Hermitian form). The practical resolution of the reducibility problem and description of the irreducible components and invariant subspaces can be very difficult. It is a consequence of complete reducibility that the problem is intimately connected with the determination of the centraliser algebra, that is, the algebra of matrices that commute with the group elements. The computation is connected with the solution of the system of linear equations $Xg = gX$, where g runs over G. For low matrix degrees, it is not hard to give an explicit solution of the system; for higher degrees and in situations that are more general, this approach is unacceptable. It is useful to have general information about the centraliser algebra $C(G)$. If the ground field is $\mathbb{C}$, then $C(G)$ is isomorphic to the direct sum of complete matrix algebras over $\mathbb{C}$ of suitable degrees. More precisely, in some basis $C(G)$ takes block-diagonal form

$$\mathrm{diag}(\overbrace{a_1, \ldots, a_1}^{r_1}, \overbrace{a_2, \ldots, a_2}^{r_2}, \ldots, \overbrace{a_s, \ldots, a_s}^{r_s}),$$

where the a_i run independently over the complex matrix algebras of suitable degrees k_i ($i = 1, \ldots, s$). It is not hard to see that the second centraliser $C(C(G))$ reduces to block-diagonal form, but with the roles of k_i and r_i interchanged. It follows from this that there are exactly s irreducible components to G (up to equivalence of the corresponding representations), and each has multiplicity k_i and degree r_i. In other words, G is similar to a group of matrices of the form

$$\mathrm{diag}(\underbrace{\varphi_1(g), \ldots, \varphi_1(g)}_{k_1}, \ldots, \underbrace{\varphi_s(g), \ldots, \varphi_s(g)}_{k_s}),$$

in which $g \to \phi_i(g)$ is an irreducible representation of G of degree r_i. It follows in particular that dim $C(G)$ is the sum of the squares of the multiplicities of the irreducible components ϕ_i of G. In the case where the identitification representation $G \to G$ is induced from an irreducible representation of a subgroup H, there is a remarkable formula due to Mackey, *viz* dim $C(G) = |H \backslash G/H|$, where the symbol on the right denotes the number of double cosets of G modulo H. It is clear that $C(G)$ is commutative if and only if $k_1 = \cdots = k_s = 1$, that is, all the ϕ_i occur with multiplicity 1, and in this case dim $C(G)$ is the number s of irreducible components ϕ_i.

The whole of the material in this section remains valid for any algebraically closed field if the order of G is relatively prime to the characteristic of the field. In the modular case, that is, when $|G|$ is a multiple of characteristic p and p is positive, questions of complete reducibility are solved only with great difficulty. There are essentially no general conditions known that would guarantee complete reducibility in the modular case. The next result is perhaps the only useful condition of this sort: if ϕ is an irreducible component (composition factor) of G whose dimension is a multiple of the order of a Sylow p-subgroup of G, then ϕ is distinguished by a direct summand.

It should be noted that the sum of the elements in a conjugacy class of elements lies in $C(G)$. It follows that a group G is reducible if class sums are not scalar matrices. In addition, if for a pair of irreducible components of G, some class sum gives different scalars, then G is decomposable. More significant decomposability conditions are known only for very special classes of linear groups.

The following classical result (Itô; see [Curtis and Reiner (1962), 53.18]) can be viewed as a reducibility condition.

Theorem. *Let G be a finite subgroup of* $\mathrm{GL}_n(\mathbb{C})$. *If n does not divide the index of any abelian normal subgroup of G, then G is reducible.*

In particular, G is reducible if the order of G is not divisible by n.

Let P be a field of positive characteristic p, and G a subgroup of $\mathrm{GL}_n(P)$. If G is known to have a normal p-subgroup H, informaton can be gained about the G-submodules of the space V_n on which $\mathrm{GL}_n(P)$ is acting. By Corollary B in 2.2.1, H reduces to triangular form in which each matrix has unit diagonal. It follows that the elements $h - E_n$ ($h \in H$) lie in the ring of upper-triangular matrices with zero diagonal. Let R be the P-linear span of these matrices $h - E_n$. Clearly, with R^i denoting the ideal generated by the elements r^1 for $r \in R$, the chain $R \supset R^2 \supset \cdots$ breaks off at some finite point k, and $R^k = 0$. Moreover, the subspaces $RV \supset R^2V \supset \cdots \supset R^{k-1}V \neq 0$ are G-modules.

4.3. Realising Field of a Finite Linear Group. In many applications, it is of interest to know whether replacing a subgroup G of $\mathrm{GL}_n(\mathbb{C})$ by xGx^{-1}, $x \in \mathrm{GL}_n(\mathbb{C})$, can produce a realisation over $\mathbb{Q}$, $\mathbb{R}$ or some other important field. A necessary condition is that the traces of the matrices in G must lie in the corresponding field. Since the eigenvalues of the matrices in a finite linear group are roots of unity, the traces lie in $\mathbb{Q}(\varepsilon)$, where ε is a root of unity, of degree d say. Let e be the least common multiple of the orders of the cyclic subgroups of G, that is, the exponent of G. Brauer (1945) showed that G is similar to a group of matrices over the field $L = \mathbb{Q}(\varepsilon)$, where ε is a primitive e-th root of unity. This result is a corollary of a deep induction theorem of Brauer in the representation theory of finite groups (see [Curtis and Reiner (1962), §41]). However, it often happens that G can be realised over a proper subfield of L. Representation theory gives some further indications on this head; for example, G can be realised over $\mathbb{R}$ if every element g of G is conjugate to g^{-1}. Similarly, G can be realised over $\mathbb{Q}$ if every element g of G is conjugate to g^k whenever their orders are the same. The symmetric group S_n satisfies this condition for every n. Thus, all its representations are realised over $\mathbb{Q}$. We note further that G can be realised over $\mathbb{R}$ if $\Phi_G = 1$ (see 4.1).

4.4. Reduction of Complex Groups to Finite Characteristic. An important device in the representation theory of finite groups and in the theory of finite linear groups is the method of reducing complex groups modulo p, where p is a prime. The questions concerns homomorphisms of a subgroup G of $\mathrm{GL}_n(\mathbb{C})$ into $\mathrm{GL}_n(P)$, where P is an algebraically closed field of positive characteristic p. Let L_p be the completion of the field L from 4.3 in the p-adic valuation (or even any locally compact field containing L such that the valuation of the element p is less than 1: see 1.4.4). Let R be the valuation ring (see 1.4.4) of L_p. By 1.4.5, a finite subgroup G of $\mathrm{GL}_n(L_p)$ is similar to a subgroup of $\mathrm{GL}_n(R)$. Comparison of this with (4.3) gives the following theorem:

Theorem A. *Let p be any prime number. Every finite subgroup G of $\mathrm{GL}_n(\mathbb{C})$ is similar to a subgroup G_1 whose matrix entries lie in the valuation ring R of some locally compact field with non-Archimedean valuation, whose field of residues is finite and has characteristic p.*

Note that the finiteness of the field of residues follows since L_p is locally compact. Let I be the valuation ideal, and set $F = R/I$. Let $R \to F$ be the natural ring homomorphism and ρ: $\mathrm{GL}_n(R) \to \mathrm{GL}_n(F)$ the corresponding linear group homomorphism. We shall call its restriction to G a reduction homomorphism and denote it by ρ_G. We emphasise that the reduction homomorphism depends not only on G, but also on its situation in $\mathrm{GL}_n(R)$. The position is this: if G_1 and G_2 are subgroups of $\mathrm{GL}_n(R)$ and $G_1 = xG_2x^{-1}$ for some $x \in \mathrm{GL}_n(L_p)$, there need not be any element y of $\mathrm{GL}_n(R)$ such that $G_1 = yG_2y^{-1}$. It can happen that $\rho_G(G)$ is completely reducible relative to one reduction but not relative to some other reduction. Moreover, the kernel Ker ρ_G of the reduction homomorphism, and its image, depend on the choice of ρ_G. We record the following simple but important fact in this context:

The kernel of a reduction homomorphism is a normal p-subgroup of G.

Set $\bar{G} = \rho_G(G)$. There is no effective way of determining the number of irreducible components of $\bar{G}$, even if G is irreducible. Note that the numbers of irreducible components of G and $\bar{G}$ are the same if $|G|$ is not divisible by p. If G

is irreducible and n is a multiple of the order of the Sylow p-subgroup of G, $\overline{G}$ is also irreducible. The following theorem, due to Brauer and Nesbitt (1937), is a key result.

Theorem B. *For fixed p, the set of composition factors of $\overline{G}$ (with multiplicities) is independent of the reduction homomorphism.*

If g is an element of G whose order is not a multiple of p, the eigenvalues of g and $\bar{g}$ are essentially the same. More exactly, if we fix some isomorphism λ from the group of $|G|$-th roots of 1 in $\mathbb{C}$ to that in P, the spectrum of $\bar{g}$ is the image of the spectrum of g under λ. The only thing that can be said about the other elements of G is that the degree of the minimum polynomial of $\bar{g}$ does not exceed that of the minimum polynomial for g, for each g in G.

We come now to an example of the use of reduction modulo p to studying the structure of a subgroup G of $\mathrm{GL}_n(\mathbb{C})$.

Theorem C. *Let g be an element of G of order p^α for some $\alpha \in \mathbb{N}$. Assume that $n \leqslant p^{\alpha-1}$. Then, G has a nontrivial normal p-subgroup.*

Proof. Let K be the kernel of the reduction homomorphism modulo p. We know that K is a p-group, and claim that $|K| \neq 1$. Reduce $\bar{g}$ to Jordan normal form. Since $(\bar{g} - E_n)^n = 0$, we have $\bar{g}^{p^\beta} = E_n$, where β is the smallest natural number such that $n \leqslant p^\beta$. Thus, g^{p^β} is a nontrivial element of K.

Theorem B has an important corollary: if P is an algebraically closed field of prime characteristic p and $(p, |G|) = 1$, the representation theory of G over P is practically identical with its representation theory over $\mathbb{C}$.

4.5. Classification of Finite Linear Groups of Small Degree over $\mathbb{C}$. It is completely clear that the classification of finite linear groups is impossible without further restrictions. One of the most convenient extra conditions to impose is to bound the degrees of the matrices involved. In many arguments using induction on the matrix degree, it turns out to be necessary to classify linear groups of comparatively small degrees.

Since finite linear groups over $\mathbb{C}$ are completely reducible, the problem comes down almost entirely to the classification of the irreducible groups. Moreover, to a significant, if not exhaustive, extent it reduces to primitive groups and then to absolutely primitive groups (see § 6 of Chapter 2, where a general strategy for this type of reduction is covered). In this section, we provide a list of the finite quasisimple irreducible groups of small degree. We are forced to restrict the description up to isomorphism, since an explicit construction of every group arising as a group of matrices over $\mathbb{C}$ would take up too much space. We borrow the lists from Feit (1970) and another paper of Feit, published in 1976. The alternating group is denoted by A_r; if H is simple, kH denotes a covering group with centre of order k:

$n = 2$: $2A_5$.

$n = 3$: A_5, $3A_6$, $\mathrm{PSL}_2(7) \cong \mathrm{SL}_3(2)$.

$n = 4$: $A_5, 2A_5, 2A_6, 2A_7, \mathrm{SL}_2(7), 6\mathrm{PSp}_4(3)$.

$n = 5$: $A_5, A_6, \mathrm{PSL}_2(11), \mathrm{PSp}_4(3)$.

$n = 6$: $2A_5, 3A_6, 6A_6, 3A_7, 6A_7, \mathrm{SL}_2(7), A_7, \mathrm{PSL}_2(7), \mathrm{SL}_2(11), \mathrm{SL}_2(13)$, $\mathrm{PSp}_4(3), \mathrm{SU}_3(3^2), 6\mathrm{PSU}_4(3^2)$, 2HaJ, $6\mathrm{PSL}_3(4)$.

$n = 7$: $\mathrm{PSL}_2(7), \mathrm{PSL}_2(13), \mathrm{PSL}_2(8), A_8, \mathrm{PSU}_3(3^2), \mathrm{Sp}_6(2)$.

$n = 8$: $A_6, 2A_8, A_9, 2A_9, \mathrm{SL}_2(8), \mathrm{SL}_2(17), 2\mathrm{Sp}_6(2), 2\Omega_8^+(3)$.

$n = 9$: $A_6, 3A_6, A_{10}, \mathrm{PSL}_2(17), \mathrm{PSL}_2(19)$.

Here HaJ denotes the Hall-Janko sporadic simple group of order $2^7 \cdot 3^3 \cdot 5^2 \cdot 7$, and $\Omega_8^+(3)$ is the commutator subgroup of the orthogonal group $O_8^+(3)$ corresponding to the quadratic form $x_1^2 + \cdots + x_8^2$.

These results are obtained without the use of the classification of the finite simple groups. Of course, using the classification, one can extend the values of n without too much difficulty and get a description of the irreducible finite groups. This has not been done explicitly as yet, as far as we know.

There are some investigations concerning the description of finite linear groups of small degree over $\mathbb{Q}$. This question is motivated by applications to describing (up to integral similarity) finite subgroups of $\mathrm{GL}_n(\mathbb{Z})$: see 6.1 below. The connection between these two questions is based on the following remarkable fact: every finite subgroup G of $\mathrm{GL}_n(\mathbb{Q})$ is similar to a subgroup of $\mathrm{GL}_n(\mathbb{Z})$. It should be noted that there exists no generalisation of this result to finite subgroups of an arbitrary (even a cyclotomic) algebraic number field, with $\mathbb{Z}$ replaced by the ring of integers of the field. We record another general result: the orders of finite subgroups G of $\mathrm{GL}_n(\mathbb{Q})$ are bounded by a quantity $d(n)$ depending only on n.

4.6. Some Applications of the Classification of Finite Simple Groups. The announcement that the classification of finite simple groups is now complete has had a profound influence on the philosophy of finite linear group theory – despite the fact that the last stages of the proof, that dealing with the classification of the so-called quasi-thin groups, has not yet been published in full. In particular, it has become clear that the solution of most of the really difficult problems using the classification theorem becomes reasonably accessible, whereas without it, the usual situation is that only very special cases have yielded to attack. All this has led to a slump in activity on problems of describing and recognising finite linear groups by methods avoiding the classification. The content of this circle of investigations, as it was prior to the announcement of the classification theorem, is fairly fully covered in the author's survey [Zalesskij (1981), Chapter 3]. By now it is apparent that a first plan is to pose the question of describing the representations with various important properties, in the first instance for simple and quasisimple groups. Based on this, there is a hope that use of the classification will lead to a description of the finite linear groups with similar properties. On the whole, activity of this sort is in a rudimentary state; however, there are some interesting results. Here are some examples.

Theorem A (Landázuri and Seitz, 1974). *Let G be an irreducible subgroup of* $GL_n(\mathbb{C})$, *where* $n > 2$ *is prime. If G is isomorphic to a (quasisimple) Chevalley group, then G is one of the following groups*:

(a) $PSL_2(q)$, $n \in \{q, q-1, q+1, (q+1)/2, (q-1)/2\}$.
(b) $GL_m(2)$, $n = 2^{m-1} - 1$.
(c) $PSL_m(q)$, $n = (q^m - 1)/(q - 1)$.
(d) $PSp_{2m}(q)$, $n = (q^m \mp 1)/2$, q *odd*.
(e) $Sp_6(2)$, $n = 7$.
(f) $PSU_m(q)$, m *odd and* $n = (q^n + 1)/(q + 1)$.

Note that this list is excessive: it can be shown that case (b) holds only for $m = 3, 4$. All the other cases can happen when the expression for n gives a prime.

The foundation for the proof of Theorem A is Landázuri and Seitz's set of lower bounds for the degrees of the irreducible complex representations of the Chevalley groups, together with the fact that n is a prime divisor of the group order (more exactly, of the order of the central factor-group: see 4.2). In fact, for all finite quasisimple groups, the degrees of the nontrivial complex representations (with rare exceptions) exceed the largest simple divisor of the group order. For example, on comparing these two quantities in tables in Conwey et al. (1985), we get without difficulty the following list of sporadic groups having representations of prime degree n: the Mathieu groups M_{11} and M_{12} with $n = 11$; M_{24} with $n = 23$, and the two Conway groups Co_{23} and Co_{24} with $n = 23$. For the alternating group A_m, the smallest degree of a nontrivial representation over $\mathbb{C}$ is known to be $m - 1$, so that m can take values n and $n + 1$. The second possibility always arises. Using more specialised properties of the representations of A_m, it is easy to see that for $m > 5$, A_m has no irreducible representations of prime degree n. Finally, we note that a quasisimple irreducible group of matrices of degree $n > 3$ is simple, since the prime divisors of the orders of the Schur multipliers of the known simple groups are not more than 3. Thus, using the classification of finite simple groups, we get a complete list of the quasisimple groups of prime degree.

In fact, Landázuri and Seitz have obtained convenient bounds for the representations of the Chevalley groups over fields of characteristic different from that of their fields of definition, as well as for the complex representations. Because of this, it has proved possible to apply these results to groups over finite fields as well as to complex linear groups. For example, Hering (1985) has described the linear groups over finite fields that contain irreducible cyclic groups of prime order. Of course, the classification of simple groups was used for this. In arguments of this type, it is impossible to avoid using the representation theory of the Chevalley groups over fields with characteristic the same as that of the field of definition of the groups concerned: the theory of modular representations of this type is incomparably better developed than that of the Chevalley groups over fields of other characteristics.

From a geometrical point of view, it is very natural to ask for a description of the linear groups G contained in $\mathrm{GL}(V) \cong \mathrm{GL}_n(q)$ acting transitively on $V_n - \{0\}$ or on the projective space $P(V_n)$. The two questions are intimately related; for example, we get groups transitive on points when scalars are added to groups transitive on lines (some groups merge in this process). The problem arose in permutation group theory as a necessary part of the classification of doubly transitive groups. This is the situation: with V_n^+ denoting the additive group of V_n, the semidirect product $G \rightthreetimes V_n^+$ acts doubly transitively on V_n if G acts transitively on $V_n - \{0\}$. The soluble doubly transitive groups were described by Huppert in 1957; on the basis of the classification of finite simple groups, Hering determined the insoluble doubly transitive groups (1985). The problem of describing permutation groups of rank 3 is of great importance in permutation group theory, and it leads to the problem of finding the linear groups having exactly two orbits on $V - \{0\}$. This problem was solved in its entirety by Liebeck in 1987. In the process, the double transitive groups were characterised anew. The bounds of Landázuri and Seitz recalled above play a significant part in the proof. In this situation, the number $a = |P(V_n)|$ divides $|G|$; comparison of the order of the Chevalley group $Chev(r)$ such that $(d, q) \neq 1$ and $(a, |Chev(r)|) = a$ with the Landázuri-Seitz number excludes almost all possibilities. A similar phenomenon occurs in the case $(r, q) = 1$. For example, if $r = q$ and $G = \mathrm{SL}_m(q)$, the equality $(a, |G|) = a$ gives that $a \leqslant m(m - 1)/2 + 1$, so that it is enough to consider the representations of $\mathrm{SL}_m(q)$ with degrees at most $m(m - 1)/2 + 1$. For $m > 5$, there are always four such representations, and analysis of these is not at all difficult.

Using the methods of the modular representation theory of finite groups, Brauer and Feit (1966) obtained the following analogue of Jordan's theorem (see 2.3.1) for fields of positive characteristic p.

Theorem B. *There exists a function $f_p(m, n)$ such that, if G is a finite subgroup of $\mathrm{GL}_n(P)$ of order not divisible by p^m, then G has an abelian normal subgroup of index not more than $f_p(m, n)$.*

Estimates for this function can be deduced from the paper of Brauer and Feit. Use of the classification theorem yields new, enormously more exact bounds for the Jordan function and the Brauer-Feit function (Weisfeiler 1984).

4.7. Finite Linear Groups of Small Degree over Fields of Prime Characteristic. As can be seen from the example discussed at the beginning of this section, the list of linear groups of small degree over fields of positive characteristic is immeasurably wider than is the case for characteristic 0. A greater complexity arises also because of the absence of complete reducibility: to describe the reducible groups, it is nowhere near enough to know the irreducible components. In this context, the problem of determining the p-groups of small degree over fields of characteristic p is fairly critical. There is no difficulty when $n = 2$ or 3, but much laborious effort is needed for larger values of n. An estimate of the size of the task could be obtained by looking at the cases $n = 4$, 5 and 6; as far as we know, nobody has attempted to do this as yet.

For small n, there is a comparatively effective procedure (described in 2.2.3) for reducing the determination of the irreducible subgroups G of $GL_n(P)$ to that of the quasisimple subgroups. However, it seems that an explicit description of all irreducible subgroups exists only in some special cases.

By now there is explicit description of the irreducible finite quasisimple subgroups G of $GL_n(P)$ with $n < 28$; for $n > 5$, the classification of finite simple groups has been used to do this. We provide here a list of the irreducible finite quasisimple linear groups of small degree, up to isomorphism. They were discovered at the beginning of the century when $n = 2$ or 3; the cases $n = 4$, 5 were considered in the seventies without the use of the classification, the tool there being some contemporary results on the recognition of finite simple groups from knowledge of their Sylow p-subgroups (see the survey Feit (1970)). The cases $5 < n < 11$ were done by Kleidman in his thesis (London University, 1986), and the cases $n < 28$ is considered by Kondratiev (Preprint, 1989).

$n = 2$: $G = SL_2(p^a)$, $SL_2(5)$, $a \in \mathbb{N}$.

$n = 3$: $G = SL_3(p^a)$, $SU_3(p^{2a})$, $PSL_2(p^a)$ with $p^a > 3$, $3A_7$ with $p = 5$, $3A_6$ with $p \neq 3$, A_5 with $p \neq 2$.

$n = 4$: $G = SL_4(p^a)$, $SU_4(p^{2a})$, $Sp_4(p^a)$, $2A_7$ with $p \neq 2$, A_7 with $p = 2$, $6PSp_4(3)$ with $p \neq 2, 3$, $4PSL_3(4)$ with $p = 3$, $SL_2(p^a)$ with $p^a > 3$, $Sz(2^{2a+1})$ with $p = 2$, $2A_6$ with $p > 2$, $2A_5$ with $p > 2$, A_5, A_6 with $p = 2$.

$n = 5$: $G = SL_5(p^a)$, $SU_5(p^{2a})$, $SO_5(p^a)$ with $p > 2$, $PSL_2(p^a)$ with $p > 3$, $PSL_2(11)$, $PSp_4(3)$ with $p > 2$, A_7 with $p = 7$, A_5 with $p > 3$, A_6 with $p > 3$, M_{11} with $p = 3$.

$n = 6$: $SL_2(p^a)$, $p^a > 5$, $G_2(2^a)$ with $p = 2$, $a > 1$, A_5 with $p \neq 2, 5$, $SL_2(7)$ with $p > 2$, $SL_2(13)$ with $p > 2$, $PSL_2(13)$ with $p = 2$, A_7 with $p \neq 7$, 2HaJ with $p > 2$, HaJ with $p = 2$, $SU_3(3^2)$, $Sp_6(p^a)$, $SL_2(11)$ with $p > 2$, $PSp_4(3)$ with $p \neq 3$, $6PSU_4(3^2)$ with $p > 3$, $3PSU_4(3^2)$ with $p = 2$, $6PSL_3(4)$ with $p > 3$, $2PSL_3(4)$ with $p = 3$, $3M_{22}$ with $p = 2$, $SU_6(p^{2a})$, $SL_6(p^a)$.

$n = 7$: $SU_7(p^{2a})$, $SO_7(p^a)$ with $p > 2$, $G_2(p^a)$ with $p > 2$, $SL_7(p^a)$, $PSU_3(3^2)$ with $p > 2$, $Sp_6(2)$ with $p > 2$, $PSL_2(p^a)$ with $p > 5$, A_9 with $p = 3$, Ja with $p = 11$, $PSL_2(13)$ with $p > 2$, $PSL_2(8)$ with $p > 2$, $PSL_2(7)$ with $p > 2$, A_8 with $p > 2$.

We restrict ourselves to this list. The notation is just as in 4.5. In addition, $\mathrm{Sz}(2^{2\alpha+1}) \cong {}^2B_2(2^\alpha)$ is a Suzuki group, M_{11} and M_{12} are the Mathieu groups, Ja is the first Janko group of order $2^3 \cdot 3 \cdot 5 \cdot 7 \cdot 11 \cdot 19$.

4.8. The Hall-Higman Theorem and Related Questions. The paper of Hall and Higman (1956) contains a result about finite linear groups which plays an important role in finite group theory as a whole. For this reason, the situation considered by Hall and Higman went on to receive considerable attention. Consequently, several versions of the proof has appeared and there have been some generalisations. We consider here the basic situation, without any pretence of maximal generality.

Theorem A. *Let G be a finite subgroup of $\mathrm{GL}_n(P)$ containing an irreducible p-subgroup A with scalar commutator subgroup. Let g be an element of order r^m in G, where r is a prime other than p, and $m > 0$. Then, the degree of the minimum polynomial of g is r^m or $r^m - 1$, the latter case occurring only when $r^m - 1$ is a power of p.*

This result was proved in the Hall-Higman paper for the case $r = \mathrm{char}\, P$. The case $r \neq \mathrm{char}\, P$ was considered by Shult in 1965. The nature of g in the exceptional case can be described more exactly. As in 2.2.3, we discover that there is a homomorphism $\theta\colon G \to \mathrm{Sp}_{2l}(p)$ with kernel AS_G, where $S_G = S \cap G$ is the group of scalar matrices in G. Then, the degree of the minimum polynomial of g in Theorem A is $r^m - 1$ if and only if $\theta(g) = \mathrm{diag}(E_k, t)$, where $0 \leqslant k < l$ and $t \in \mathrm{Sp}_{2(l-k)}(p)$ is an element of order $p^{l-k} + 1$ (a "Singer cycle"). This provides a reason for a more detailed consideration of the normaliser N of A in $\mathrm{GL}_n(P)$. Note that the irreducibility of A implies that $n = p^l$. We have the isomorphism $N/AS \cong Sp_{2l}(p)$ (see [D.A. Suprunenko (1972), § 20, Theorem 16]), except when $p = 2$ and S_G has no elements of order 4. (In that case $N/AS \cong O^{\mp}_{2l}(2)$.) This enables one to use properties of $\mathrm{Sp}_{2l}(p)$ in analysing N and its elements. For example, one can study the case $r = p \neq \mathrm{char}(P)$; for $m = 1$, $r > 2$, it turns out that the degree of the minimal polynomial is less than r only when $\theta(g)$ is a transvection (Zalesskij (1986); the case $m > 1$, $r = p > 2$ has been studied by Beglarian and Zalesskij (1991)).

Suppose that $p > 2$. In that case, N splits: $N = AS_G \rtimes H$, where $H \cong \mathrm{Sp}_{2l}(p)$. At the same time, one gets a realisation of $Sp_{2l}(p)$ by matrices of degree p^l over P, called the Weil representation. (Weil (1964) considered the p-adic version of this construction and representations in Hilbert space; it has important applications in number theory. This realisation of $\mathrm{Sp}_{2l}(p)$ was studied by Bolt, Room and Wall (1961).) The restriction of this representation to $\mathrm{GL}_n(p)$ and $U_n(p)$ are called the Weil representations of these groups. Gérardin (1977) computed the character of the Weil representation. It turns out that the Weil representation of $\mathrm{GL}_n(p)$ is "almost" equivalent to a permutation representation. More exactly, consider the action of this group by permutations on the points of an n-dimensional vector space V_n over the field of p elements. Let W be a vector space of dimension $|V_n|$ over P with basis indexed by the elements of V_n. The action of

$GL_n(p)$ on V_n induces a corresponding action on W, resulting in a representation $\pi: GL_n(p) \to GL(W)$, which we shall call the permutation representation. It is not hard to check that the character χ of this representation is given by the formula $\chi(g) = p^{f(g)}$, $g \in GL_n(p)$, where $f(g)$ is the dimension of the stabiliser of g in V_n. Let δ be the unique nontrivial representation of $GL_n(p)$ (and of $U_n(p)$) in $\{\mp 1\} \subseteq P$ if $p > 2$ and char $P \neq 2$; otherwise, set $\delta = 1$. Then, the Weil representation of $GL_n(p)$ is of the form $\pi \otimes \delta$, and its character is $\chi\delta$. In a sense, the character of the Weil representation of $U_n(p)$ is dual to that of $GL_n(p)$. According to Gérardin, it is of the form $\delta(g)\cdot(-1)^n(-p)^{f(g)}$, where $f(g)$ is the dimension over $\mathbb{F}_{p^2}$ of the stabiliser of g in a vector space of dimension n over $\mathbb{F}_p 2$, and we are given the natural representation of $U_n(p)$.

When $p = 2$, the factor-group $Sp_{2l}(2)$ does not split off from N for $l > 3$. This follows because 2^l is less than the degree μ of the minimal representation of $Sp_{2l}(2)$, which is bounded by the number $2^{l-2}(2^{l-1} - 1)$, and that is at most μ (Landázuri and Seitz, 4.6). At the same time, this shows that the extension N of the group $H = Sp_{2l}(2)$ can have a representation of degree less than that of H itself while being faithful on H. The Landázuri-Seitz bounds also show that the subgroups $O_{2l}^{\mp}(2)$ of $Sp_{2l}(2)$, as well as the subgroups $Sp_{2r}(2^t)$ of $O_{2r}^{\mp}(2^t)$ in the case $l = rt$, furnish similar examples. In 1978, Feit and Tits showed that for this list of groups H, the number 2^l is the smallest d such that there exists an immersion of H in $GL_d(\mathbb{C})$ as a "section", that is, as a representation of some extension. We note further that something similar holds for the Chevalley series $G_2(2^t)$, for even $t > 2$.[4] The situation is that there is an embedding $G_2(2^t) \subset Sp_6(2^t) \subset Sp_{6t}(2)$ such that $G_2(2^t)$ is a section of $GL_{2^{3t}}(\mathbb{C})$. On the other hand, the character table of this group (Enomoto and Yamada 1986) shows that the smallest degree of a complex representation is $2^{3t} + 1$ for $t > 2$. There is a connection between this problem and the important question of determining the non-split extensions $R \lhd X$, where R is an elementary abelian group and $Y \cong X/R$ is a finite simple group acting irreducibly on R (that is, we are given a representation $Y \to GL(R)$). Very little is known here. It follows from the above remarks that, for $p = 2$, there exist non-split extensions with $R = A/S_G$ of $Sp_{2r}(2^t)$, $O_{2r}^{\mp}(2^t)$, $G_2(2^t)$. As far as we are aware, the uniqueness problem for these extensions is still open.

4.9. Eigenvalues of the Matrices in Finite Linear Groups. A wide cycle of investigations on this theme was undertaken by Blichfeldt at the beginning of the century; a large proportion of his results appear in his book published in 1917. The following is one of the most useful of these for very wide application in the classificaton of finite linear groups over $\mathbb{C}$.

Theorem A. *Let G be a finite primitive irreducible subgroup of $GL_n(\mathbb{C})$, and g a non-scalar matrix in G. Then, the set of eigenvalues of g is not contained strictly inside any 60°-sector of the unit circle in the complex plane.*

---

[4] Kleidman and Liebeck (1989) have shown that for all other simple groups H other than $PSL_4(2^m)$, $m > 2$, d is the smallest degree of a projective representation *of* H.

For any g in $\mathrm{GL}_n(\mathbb{C})$, we let $m(g)$ stand for the order of g modulo the group S of scalar matices, so that $m(g)$ is the order of the image of g in the projective linear group $PGL_n(\mathbb{C})$. Another valuable result of Blichfeldt's is this: for $n > 1$, a primitive group G cannot contain a matrix g with $m(g) > 5$ and having two different eigenvalues (not including multiplicities). This is not true when $m(g) = 5$. However, it can be shown in this case that G contains a subnormal subgroup isomorphic to $\mathrm{SL}_2(5)$ (see [Zalesskij (1981), § 11]). When $m(g) = 4$, G must contain a non-scalar normal 2-subgroup; this follows from arguments similar to those used in the proof of Theorem C in 4.4 for $p = 2$. The case $m(g) = 3$ has not been considered, but will not be very hard if the classification of finite simple groups is used. We note that when $m(g) = 3$, there are some fairly complicated groups arising; for example, $\mathrm{Sp}_{2n}(3)$ in its representation of degree $(3^n - 1)/2$ over $\mathbb{C}$. Clearly, the case $m(g) = 2$ cannot have a reasonable outcome. We must also note that every group G generated by elements g with $m(g) = 5$ is a Kronecker product of groups isomorphic to $\mathrm{SL}_2(5)$. If G is generated by elements g with $m(g) = 4$, then G is the Kronecker product of groups H_i containing normal subgroups A_i with scalar commutator subgroups such that H_i/A_i is one of the following groups (up to isomorphism): the symmetric group S_m, the symplectic group $\mathrm{Sp}_{2n}(2)$, the orthogonal group $O^+_{2m}(2)$ or $O^-_{2m}(2)$ (Korlyukov 1986).

Properties of linear groups associated with eigenvalues are essentially geometrical in nature. The absence of 1 as an eigenvalue means that the map has no fixed points in the vector space. It is quite natural that this property is most often met in applications. In connection with his work on near-fields, Zassenhaus (1935) gave a partial characterisation of the finite linear groups over $\mathbb{C}$ with no fixed points. This is the class of finite linear groups G containing no non-identity matrix with 1 as eigenvalue. A complete description was obtained by Vincent (1947) as part of his solution of the Clifford-Klein problem about spherical space forms. Just how important Vincent's result is in geometry is illustrated in part by the fact that a significant part of the book by Wolf (1972) on spaces of constant curvature is given over to an exposition of the work of Zassenhaus and Vincent. The list of groups without fixed points is rather enormous, and we shall not give it here. We merely remark that they have very simple structure from the purely group-theoretical viewpoint.

There is an impression that almost every matrix in a perfect primitive group has 1 as an eigenvalue. Little is known in full generality. We refer at this point to the results of Shult (1965) mentioned in the previous section, and to the following result of Zalesskii (1989). Let H be a finite simple group of type E_8, F_4 or ${}^2F_4$. Then, in every irreducible complex representation of degree more than one of H, every semisimple element h has 1 as an eigenvalue. We can expect that this is so for all the elements of H, not just the semisimple elements. Note further that if the group G figuring in Theorem A of 4.8 is perfect, then 1 fails to be an eigenvalue of the matrix $g \in G$ involved only when $p = 2$.

The most inflexible regularity in the behaviour of eigenvalues is observed for simple or quasisimple finite groups. The most natural approach at present for analysing this regularity is to use the representation theory of the known simple groups, and the classificaton theorem for finite simple groups. It is expected that an element g of a quasisimple group G has exactly $m(g)$ different eigenvalues, except in a certain series of exceptions. For example, if g is an element of order p in an irreducible representation $\phi \neq 1$ of a finite quasisimple Chevalley group of characteristic p, then $\phi(g)$ has p different eigenvalues, except in the following cases:

(a) $G = SU_3(p)$, g a transvection; (b) $G = SL_2(p^2)$,

(c) $G = Sp_{2r}(p)$, $r \geqslant 1$, g a transvection;

(d) $G = Sp_4(p)$, g not a transvection. Moreover, g fails to have 1 as an eigenvalue only in cases (a), (b), (d). (Zalesskij 1986)

Of the general results, the most significant is the following theorem of Robinson (1983):

Theorem B. *Let G be a finite primitive irreducible group, $G \subset GL_n(\mathbb{C})$, and g a noncentral element of G of order p. Then, g has not less than $(p + 3)/4$ different eigenvalues.*

The proof of this theorem does not rest on the classification theorem; instead, it uses the techniques of ordinary and modular representation theory. It is probable that the limiting lower bound for the number of eigenvalues of g in Theorem B is $(p - 1)/2$. Observe that, in case (c), the number of different eigenvalues of g is $(p + 1)/2$ when $r > 1$, and $(p - 1)/2$ when $r = 1$; it is exactly $p - 1$ in cases (a), (b), (d).

The problem of finding bounds for the multiplicities of eigenvalues is an interesting one. On the whole, one must expect that the multiplicities of the eigenvalues of a matrix g in a finite irreducible primitive subgroup G of $GL_n(\mathbb{C})$ lie between $n/10m(g)$ and $10n/m(g)$. Still, the symmetric group on $n + 1$ symbols in its representation of degree n presents a sharp divergence from the expected picture. However, it is probable that this exception is unique in some sense. This time there exists a method of estimating the maximum $\nu(H)$ of the multiplicities of the eigenvalues of nonscalar matrices in a finite primitive linear group H. For complex linear groups it was discovered by J. Hall (1988) and Gordeev (1990). Later Hall, Liebeck and Seitz (preprint, 1991) developed a modular version of the method. In the modular case, dimensions of eigenspaces are used instead of multiplicities of eigenvalues. Their main result is:

Theorem C. *Let H be a finite primitive tensor-indecomposable subgroup of $GL_n(F)$ such that $\nu(H) > n^{1/2}/12$. Then H is either one of $A_{n+1}, A_{n+2}, S_{n+1}, S_{n+2}$ (alternating or symmetric groups) or a classical group of dimension n over a finite field of the same characteristic as F.*

The method is based on two observations. The first is trivial: if H is generated by k elements h_i with $n - \nu(\langle h_i \rangle) < d$, then $n < kd$. The second is this. Suppose that $H = \phi(G)$, where G is simple modulo its centre. For g in G, we let $k(g)$ stand for the minimum number of conjugates of g needed to generate G, and set $k = \max_{g \in G}(k(g))$. We define another numerical invariant $m(G)$, as follows. For $G \cong A_m$, $m(G) = m$; for classical G, $m(G)$ is the degree of the natural realization of G. Then k is bounded by a linear function of m. On the other hand, dim $\phi = n$

is bounded below by a quadratic function, unless ϕ happens to be the representation figuring in Theorem C. It follows that d is not too small in comparison with n, and hence neither is $\nu(H)$.

This theorem needs the classification of finite simple groups to prove it. The main part of the proof is an analysis of the representations of alternating and Chevalley groups. We note that no way is known of finding lower bounds for the multiplicities of eigenvalues.

§ 5. Classical Groups over Rings

Chapter 1 contained a rough exposition of the elements of the theory of classical groups over fields. A reasonably systematic coverage of the theory would entrail repeating the material in Dieudonné's classical monograph Dieudonné (1971) devoted to this theme. This is less than advisable, and we prefer to refer the interested reader directly to that work. Two tendencies in the later development of the theory of classical groups should be noted. On the one hand, the theory is a substantial part of the more general theory of algebraic groups. Elementary computations have given way to powerful and considerably more universal methods of an algebraic-geometrical nature, thus enriching the range of problems tackled by the theory. However, it should not be forgotten that the classical groups constitute a basic part of the list of simple algebraic groups. On the other hand, the theory of the classical groups over noncommutative rings of a fairly general type has been intensively developed in recent decades. This material cannot be included in any form in the theory of algebraic groups. The spirit of the problems that develop here is nearer to those in ring theory; it is entirely possible that in time the theory will become part of that of the multiplicative groups of rings, in which all that remains of the matrices is certain systems of orthogonal idempotents. Nonetheless, currently this theme is being developed in the main as a branch of the theory of linear groups. The machinery produced here is essentially a part of that of algebraic K-theory (see Bass (1968) et al.)

Classical groups over rings is the discipline that inherits in the greatest degree the traditional ideology of the theory of the classical groups. The main questions here are those of describing normal subgroups and automorphisms. The analysis of systems of generators and relations also has important influence. The present-day tendency is to seek results for non-commutative rings of the widest possible nature. To our way of thinking, the theory has achieved a high degree of perfection in recent years.

5.1. Structure of the General Linear Group over a Ring. Let R be an associative ring with identity, and I a two-sided ideal of R. The elementary subgroup $\mathrm{EL}_n(R)$ has an important part to play in the theory of linear groups over rings.

By definition, it is generated by the elementary matrices, that is, the matrices $E_n + xe_{ij}$, for $x \in R$, $1 \leqslant i, j \leqslant n$, $i \neq j$. If R is a field, then $\mathrm{EL}_n(R)$ is simply $\mathrm{SL}_n(R)$. The elementary matrices satisfy relations of a fairly simple form (see 5.4 below); because of this, $\mathrm{EL}_n(R)$ is very amenable to study from a group-theoretical standpoint. However, its influence on the structure of $\mathrm{GL}_n(R)$ is not completely clear in full generality. From this point of view, the role of $\mathrm{EL}_n(R)$ seems to be more important if it is normal in $\mathrm{GL}_n(R)$. The basic fact is that this happens when $n > 2$ and R is any commutative ring (Suslin 1977). It is not always true when $n = 2$. If R is not commutative, $\mathrm{EL}_n(R)$ need not be normal in $\mathrm{GL}_n(R)$ even when $n > 2$ (Gerasimov 1987). Let us give some futher explanation at this point. For every field P, natural number n, and nontrivial group G, there exists a ring $R = R(P, n, G)$ such that $M_n(R)$ is the free product of the matrix ring $M_n(P)$ and the group ring PG, with P amalgamated. If P is of characteristic 0, then $\mathrm{GL}_n(R)$ is the generalised free product of $\langle D_n(R), EL_n(R)\rangle$ and some group H containing P^* with P^* amalgamated (P^* is the multiplicative group of P). Here, $D_n(R)$ is the group of diagonal matrices over R; clearly, it normalises $\mathrm{EL}_n(R)$. If G is an ordered group (for example, if it is free, or torsionfree nilpotent), the group of invertible elements of PG is just P^*G; in this case, H is isomorphic to P^*G. However, H is always different from P^*. It follows from the general theory of free products with amalgamation that $\mathrm{EL}_n(R)$ is not normal (or even subnormal) in $\mathrm{GL}_n(R)$.

To describe results on the structure of normal subgroups of $\mathrm{GL}_n(R)$, we need to introduce certain subgroups connected with ideals of R. Let $\mathrm{GL}_n(R, I)$ be the subgroup consisting of the matrices in $\mathrm{GL}_n(R)$ congruent to E_n modulo the two-sided ideal I of R. Further, let $C_n(R, I)$ be the subgroup consisting of all the matrices x in $\mathrm{GL}_n(R)$ such that $xa \equiv ax \pmod I$ for every a in $\mathrm{GL}_n(R)$. Also important is the subgroup $\mathrm{EL}_n(R, I)$, defined to be the normal closure in $\mathrm{EL}_n(R)$ of the set of matrices $E_n + xe_{ij}$, where $x \in I$, $1 \leqslant i, j \leqslant n$, $i \neq j$. If R is commutative and $n > 2$, $EL_n(R, I)$ is a normal subgroup of $\mathrm{GL}_n(R)$ (Suslin 1977). Finally, $C_n(R, I)$ is normal in $\mathrm{GL}_n(R)$ for every R. We shall agree to say that, for $n > 2$, the description of the normal subgroups of $\mathrm{GL}_n(R)$ is standard if all the groups $\mathrm{EL}_n(R, I)$ are normal in $\mathrm{GL}_n(R)$, and for every normal subgroup H of $\mathrm{GL}_n(R)$ there exists an ideal I of R such that

$$EL_n(R, I) \subset H \subset C_n(R, I).$$

Questions of determining the normal subgroups of $\mathrm{GL}_n(R)$ are considered in detail in Bass (1968) in connection with applications to algebraic K-theory. In particular, a concept of stable rank is introduced there which is applicable to arbitrary R, and it is shown that the determination of the normal subgroups of $\mathrm{GL}_n(R)$, $n > 2$, is standard provided that n is greater than the stable rank of R. By now, standardness has been established for every commutative ring and $n > 2$ (J.S. Wilson 1972; Golubchik 1973). There is also a class of noncommutative rings contaning rings with polynomial identities and regular rings (in the sense of von Neumann) for which the normal subgroups of $\mathrm{GL}_n(R)$ have been shown to exhibit standard behaviour for $n > 2$ (Golubchik 1985). These are the rings with the following properties: (a) every primitive homomorphic image of R is a right Ore ring, (b) there is a natural number m and a chain $\{0\} \subset I_0 \subset I_1 \subset \cdots \subset I_m \subset I_{m+1} = R$ of ideals of R such that, for every ideal P of R/I_k contained in I_{k+1}/I_k ($k = 0, \ldots, m$), there exists a right ring of fractions of R/I_k relative to the multiplicative system $1 + P$. The above result of Gerasimov shows that the class of rings R for which normal subgroups shows standard behaviour for $n > 2$ cannot be too wide. It is less probable that there will exist a reasonable description of this class in ring-theoretical terms.

The case $n = 2$ is qualitatively different from that where $n > 2$. To be precise, the structure of the normal subgroups of $\mathrm{GL}_2(R)$ is more heavily dependent on the structure of R than when $n > 2$. Thus, if $R = \mathbb{Z}$, $\mathrm{GL}_2(\mathbb{Z})$ is close to being free, in that it contains a free subgroup of finite index. It follows from this that $\mathrm{GL}_2(\mathbb{Z})$ has hugely many subgroups of finite index. On the whole, the behaviour of $\mathrm{GL}_2(R)$ becomes more regular as R grows. The reader will find a bibliography on this subject in the survey Zalesskij (1981).

If R is commutative, it is natural to ask whether $\mathrm{SL}_n(R, I)$ and $EL_n(R, I)$ are the same. In general, they are not, indeed equality appears to be fairly rare. In some cases, it has been possible to give an exact calculation of $\mathrm{SL}_n(R, I)/\ \mathrm{EL}_n(R, I)$; the classical case is when $n > 2$ and R is the ring of integers in an algebraic number field (Bass, Milnor and Serre 1968). We emphasise that for rings of this sort the problem is solved in the wider context of the arithmetic theory of algebraic groups: see 1.6.11.

In 1976, a systematic effort was initiated to get a description of the linear groups over a ring R that contain the diagonal group (Borevich, Vavilov *et al*). The most conclusive results have been obtained for semilocal R. It is fairly clear that this question is harder the fewer invertible elements R has. To avoid this obstacle, consider the following version of the problem: describe the linear groups containing the group of block-diagonal matrices with a reasonably small set of block-dimensions. This problem has been solved in cases where all blocks are of dimension not less than 3 and R is commutative. The solution is couched in terms of net groups (see 1.1.8). Many properties of net groups are clarified along the way.

Another interesting line of investigation in this region is the problem of defining the subgroups sitting between $\mathrm{GL}_n(K)$ and $\mathrm{GL}_n(R)$, where K is a subring of R. It seems that very strong uniformity must hold here, though comparatively little has actually been done. The most important result in the area describes the intermediate subgroups in the case where K is a Dedekind ring or a Bezout ring (see Bourbaki (1961–65) for this term) and R its field of fractions (R.A. Schmidt 1981).

5.2. Structure of Classical Groups over Rings. Let K be any ring. A map $\sigma: K \to K$ is said to be an *involution* if it is an automorphism of the additive group of the ring and satisfies the conditions $\sigma(k_1 k_2) = \sigma(k_2)\sigma(k_1)$, $\sigma(\sigma(k)) = k$ for all k, k_1, k_2 in K.

Suppose that $M_n(R)$ has an involution σ. We set $U^{(\sigma)} = \{g \in \mathrm{GL}_n(R) | \sigma(g)g = E_n\}$. It is clear that $U^{(\sigma)}$ is a subgroup, and that the unitary, symplectic and orthogonal groups (over fields of characteristic not 2) can be written in the form $U^{(\sigma)}$ for suitable σ when R is a field. Further, this definition is obviously the most natural generalisation of the concepts of unitary, orthogonal and symplectic groups over a field. However, the groups $U^{(\sigma)}$ are hard to deal with in the widest generality. For this reason, it is usual to consider involutions σ of a more special form. Let R be a ring with involution $r \to \bar{r}$, $r \in R$. Then the map $g \to {}^t\bar{g}$, $g \in M_n(R)$, is an involution of the matrix ring. Moreover, if Γ is a matrix in $\mathrm{GL}_n(R)$ such that ${}^t\bar{\Gamma} = \varepsilon\Gamma$, where ε is in the centre of R and $\varepsilon\bar{\varepsilon} = 1$, the map $\sigma\colon g \to \Gamma {}^t\bar{g}\Gamma^{-1}$ is an involution of $M_n(R)$. We set $U(\Gamma, R) = U^{(\sigma)}$ for the involution σ thus defined. The group $U(\Gamma, R)$ is usually more concrete. One can introduce a concept of standard description of normal subgroups that imitates the definition for $\mathrm{GL}_n(R)$. Apparently, the most general result about normal subgroups of groups like $U^{(\sigma)}$ is due to Golubchik (1985). He showed that the description is standard when R satisfies items (a) and (b) of the preceding section, and Γ and Γ^{-1} have zeros in positions (i, j) with $1 \leqslant i, j \leqslant 3$ (or, in another version, in positions (i, j) with $1 \leqslant i, j \leqslant 2$ under the additional hypothesis that R is generated *qua* ideal by the elements $\lambda - \bar{\lambda}$, where λ runs over the centre of R).

It should be noted that almost all the problems discussed in 5.1 can be posed for classical groups over rings with involutions. A number of versions appear, and the differences between them can be qualitative in nature. For example, even in the case where R is a field, the group of diagonal matrices in $U^{(\sigma)}$ depends very heavily on the choice of Γ. These questions are considered in more detail in the survey Zalesskij (1983).

5.3. Automorphisms and Isomorphisms of Classical Groups over Rings. The problems discussed in this discipline grew out of the classical results of Dickson (1901), van der Waerden (1935) and Dieudonné (1948) about automorphisms and isomorphisms of classical groups over fields. The automorphisms form a group, and the modern approach to describing it is to give a set of generators declared to be "standard automorphisms", so that every automorphism is a product of standard automorphisms.

First of all, we consider some constructions of automorphisms of the group K^0 of invertible elements of an arbitrary ring K with identity, bearing in mind the basic case $K = M_n(R)$, $K^0 = \mathrm{GL}_n(R)$, R some ring. Clearly, every automorphism of K induces an automorphism of K^0. Note too that every anti-automorphism β of K induces an automorphism of K^0 given by the rule $x \to \beta(x^{-1})$, $x \in K^0$. If K is the direct sum of rings K_1 and K_2, then K^0 is the direct product of the groups K_1^0 and K_2^0. Thus, these two devices can be united into one: a map r that is an automorphism of K_1 and an anti-automorphism of K_2 induces an automorphism of K^0. We call an automorphism like this an automorphism of linear type. Another fundamental device for constructing an automorphism of K^0 is to use a homomorphism δ from K^0 into its centre $Z(K^0)$: the map $x \to x \cdot \delta(x)$ is an automorphism of K^0, and such things are called central automorphisms. These two types of automorphisms are those declared to the standard. We shall say that the automorphism group $\mathrm{Aut}(K^0)$ has standard description if it is generated by standard automorphisms. We can go further and try to express automorphisms of the first type in terms of simpler ones.

If the ring K has an involution σ, the group $K^{(\sigma)} = \{k \in K \mid \sigma(k)k = 1\}$ is called the *unitary group of K with respect to σ*. If α is an automorphism of K commuting with σ, then α induces an automorphism of $K^{(\sigma)}$. As above, it is proper to refer to such automorphisms as being of linear type. A homomorphism of $K^{(\sigma)}$ into its centre induces an automorphism of $K^{(\sigma)}$ *via* the above formula. Automorphisms of these types are said to be standard; if every automorphism is a product of standard automorphisms, we say that the description of the automorphism group of $K^{(\sigma)}$ is standard.

Now suppose that $K = M_n(R)$ and $K^0 = \mathrm{GL}_n(R)$. If R is commutative, the description of the automorphisms of $\mathrm{GL}_n(R)$ is standard for $n > 3$ (Waterhouse 1980), and also when $n = 3$ provided that $1/2 \in R$ (Petechuk 1982).

In the noncommutative case, the situation is very considerably distorted. The main result here is this. If $n > 2$ and $1/2 \in R$, every automorphism of $\mathrm{EL}_n(R)$ is standard, that is, it is the restriction to $\mathrm{EL}_n(R)$ of a standard automorphism of $\mathrm{GL}_n(R)$. (Golubchik and Mikhalev 1985, Zel'manov 1985). However, this result cannot be extended to $\mathrm{GL}_n(R)$ if R is arbitrary. This follows from the result of Gerasimov mentioned in 5.1.

The following wider problem is a natural one. Describe the isomorphisms between $\mathrm{GL}_n(R)$ and $\mathrm{GL}_m(S)$ for rings R and S. It has been actively investigated in recent years; essentially, results here have reached the same sort of level as in the automorphism case. Similarly, isomorphisms between other classical groups are being studied. The following result of Golubchik and Mikhalev (1983) is the most general known to the author.

Let K and σ be as above, suppose that $1/2 \in K$ and that e is an idempotent of K (so that $e^2 = e$). Suppose further that the following three conditions are satisfied: (a) there exist elements x and y of K^0 such that $\sigma(x) = -x$, $\sigma(y) = y = y^{-1}$, $\sigma(e) = xex^{-1}$, $KxK = K$, and the products in pairs of different elements from the set $\{e, \sigma(e), yey\}$ are 0; (b) $K(1 - e - \sigma(e) - y(e + \sigma(e))y)K = K$; (c) the centre Z of K^0 contains an element r such that $r\sigma(r) - 1 \in Z$. Denote by $E_e(K^{(\sigma)})$ the subgroup of $K^{(\sigma)}$ generated by the elements of the form

$$1 + eK(1 - e) - \sigma(ek(1 - e)) - \tfrac{1}{2}ek(1 - e)\sigma(ek(1 - e)),$$

$$1 + (1 - e)ke - \sigma((1 - e)ke) - \tfrac{1}{2}\sigma((1 - e)ke)\cdot(1 - e)ke, \qquad k \in K.$$

Further, let S be a ring with 1/2 and an involution τ. Suppose that there is an isomorphism $\theta\colon K^{(\sigma)} \to S^{(\tau)}$ for which $-1 \in \theta(E_e(K^{(\sigma)}))$. Then, there is a ring isomorphism ϕ from K to S such that $\phi(k) = \theta(k)$ for all $k \in E_e(K^{(\sigma)})$. In particular, if $K = S$, the automorphism θ is standard on $E_e(K^{(\sigma)})$. This result can be applied to get a description of the group $U(\Gamma, R)$ of automorphisms as in 5.2, given certain assumptions about the matrix Γ.

5.4. Generators and Relations. One of the fundamental aspects of this circle of problems is to give Chevalley groups over rings in terms of generators and relations adjusted to root systems. The classical case is that of describing the special linear group by means of defining relations between the elementary matrices. In this subsection, K is a commutative ring. Convenient notation for the elementary matrix $E_n + \alpha e_{ij}$ is e_{ij}^{α}; then $e_{ij}^{\alpha} e_{ij}^{\beta} = e_{ij}^{\alpha+\beta}$ for all α, $\beta \in K$. Denoting the commutator $aba^{-1}b^{-1}$ of two invertible matrices by (a, b), we have

$$(e_{ij}^{\alpha}, e_{jl}^{\beta}) = e_{il}^{\alpha\beta} \ (i \neq l) \quad \text{and} \quad (e_{ij}^{\alpha}, e_{kl}^{\beta}) = 1,$$

if $j \neq k$, $i \neq l$. For $n > 2$, it turns out that these relations characterise $\mathrm{EL}_n(K)$ to a considerable extent. More exactly, let $\mathrm{St}_n(K)$, $n > 2$, be the group with generators x_{ij}^{α}, $\alpha \in K$, and defining relations

$$x_{ij}^{\alpha} x_{ij}^{\beta} = x_{ij}^{\alpha+\beta}, \quad (x_{ij}^{\alpha}, x_{jl}^{\beta}) = x_{il}^{\alpha\beta} \quad \text{for} \quad i \neq l;$$

$$(x_{ij}^{\alpha}, x_{kl}^{\beta}) = 1 \quad \text{for} \quad j \neq k, \quad i \neq l.$$

It follows from general group-theoretic considerations that the map $x_{ij}^{\alpha} \to e_{ij}^{\alpha}$ extends to a group homomorphism from $\mathrm{St}_n(K)$ to $\mathrm{EL}_n(K)$.

One of the most remarkable facts in this area is that the kernel of this homomorphism is central when K is commutative and $n > 4$. Informally, this means that the above list of relations determines $\mathrm{EL}_n(K)$ almost completely. In some cases, for example when K is a finite field, the groups $\mathrm{ST}_n(K)$ and $\mathrm{EL}_n(K)$ are isomorphic, that is, our relations define $\mathrm{EL}_n(K)$ completely (remember that this group is $\mathrm{SL}_n(K)$ in the case of a field). The following result holds in general, and it plays an important part in algebraic K-theory. For $n > 4$, $\mathrm{St}_n(K)$ is the universal central extension of $\mathrm{EL}_n(K)$. We recall that a group G is the universal central extension of a group H if G is perfect, can be mapped homomorphically onto H with central kernel, and every central extension F of G splits, that is, the kernel of the corresponding homomorphism $F \to G$ is a direct factor of F.

In fact, this theory has been developed not only for $EL_n(R)$, but also for the subgroups generated by root unipotent elements in arbitrary Chevalley groups over commutative rings. The reader will find a fairly detailed exposition of this theme in Steinberg (1967), §§ 6, 7.

Other investigations of generators and relations are conducted in the main for classical groups over fields, the motivation being geometrical applications. This is the situation. If X is a group of automorphisms of some geometrical object, certain generating sets have a good geometrical interpretation and seem to be preferable to others. Provision of a generating set means the statement of a geometrical result asserting that every map is the result of performing in succession certain standard maps realising the generators. Examples are reflections in hyperplanes in orthogonal geometry. A classical theorem states that the orthogonal group is generated by reflections (see Bourbaki [(1958), Chapter IX, § 6.4]). The following is an important question in this context: is there a bound for the number of factors in minimal representations for the elements of the group in terms of the generators? For example, every orthogonal transformation of a space of dimension n is the product of at most n reflections. There are many papers containing solutions to problems of this sort for various groups and widely different generating sets. Results obtained in recent years, with the corresponding references, can be found in § 5 of Zalesskij (1983).

The problem of finding a complete system of relations between the generators seems to be more difficult, as does the determination of a minimal system of relations of which all others are consequences. In a series of articles in the years 1977–1982, Ellers studied minimal systems of relations between generators of pseudoreflection type in classical groups.

§ 6. A Survey of Some Other Topics

The theory of linear groups is by no means exhausted by the themes considered so far. The aim of this section is to give a short survey of some other directions in the theory and to provide the reader with a guide to the literature.

6.1. Integral Linear Groups. The classical groups over $\mathbb{Z}$ are considered to be part of the arithmetic theory of algebraic groups. As well as this, there are questions specific to the groups themselves. For example, from a geometric point of view, there is interest in the question about when an integral orthogonal group contains a subgroup of finite index generated by reflections.

Among the non-classical integral linear groups, it is the finite groups that have attracted most attention. The motivation is chiefly applications to high-dimensional crystallography (see [Zalesskij (1983), § 13]), as well as connections with the integral representation theory of finite groups. A complete description of the finite groups of integer matrices exists only in dimensions at most 4. In dimensions up to nine, the classification of the maximal absolutely irreducible finite subgroups of $GL_n(\mathbb{Z})$ is complete. See Zalesskij (1983) for more details about this. More recently, Plesken (1985) has described the maximal finite irreducible subgroups of $GL_n(\mathbb{Z})$ for $n = 11, 13, 17, 19, 23$.

It follows from the remarks in 4.1 that a finite subgroup G of $GL_n(\mathbb{Z})$ must preserve some positive definite quadratic form f on the free $\mathbb{Z}$-module of rank n (on which $GL_n(\mathbb{Z})$ has a natural action). On the other hand, the group $O_n(f)$ of integer matrices preserving f is finite. This follows from the fairly obvious fact that it is simultaneously discrete and compact. Therefore, the maximal finite subgroups of $GL_n(\mathbb{Z})$ appear among the subgroups of $O_n(f)$. However, there is no effective procedure for recognizing $O_n(f)$. No less difficult is finding a form f in practice when G is given and has rather large order. Note that the form is unique up to a multiplicative constant if G is absolutely irreducible. Usually, it is preferred to fix f by the requirement that $a^{-1}f$ is not an integral form for integers $a > 1$. An important question is whether f is unimodular, that is, its matrix has determinant ∓ 1. Little is known here. The Leech form on $\mathbb{Z}^{24}$ is exceptionally important in the theory of finite simple groups. Twelve of the twenty-six known sporadic groups are realized in quite definite ways in the group $O_{24}(f)$ associated with the Leech form f. Note that the Leech lattice is not unimodular.

Investigations on $O_n(f)$ are still carried out even when f is not positive definite. Recent years have seen active development of the theory of the $O_n(f)$ in the case where f has signature $(n-1, 1)$; these are the hyperbolic forms. In such cases, the $O_n(f)$ are infinite and irreducible. The subgroups generated by reflections are important in geometrical applications.

For hyperbolic f Vinberg proved in 1981 that every subgroup of $O_n(f)$ generated by reflections has infinite index for all $n > 30$. Nikulin (1981) has given a complete classification of the forms f for which $O_n(f)$ contains a subgroup of finite index generated by reflections in hyperplanes orthogonal to integral vectors of length 2.

The cohomology of certain classical groups of integer matrices, and their congruence subgroups, have been the subject of some study.

There is no diminution in the investigation of subgroups of $\mathrm{GL}_2(\mathbb{Z})$ and $\mathrm{SL}_2(\mathbb{Z})$; the motivation is – or at least, it was originally – the theory of automorphic forms. Commentary and bibliography can be found in the surveys Vol'vachev and D.A. Suprunenko (1965), Zalesskij (1983), Merzlyakov (1971). It should be noted that there is a close connection between the general theory of automorphic forms and the theory of discrete subgroups of Lie groups and arithmetic subgroups of algebraic groups; however, only for $\mathrm{GL}_2(\mathbb{Z})$ has a systematic description of the subgroups been attempted.

6.2. Linear Groups over Division Rings. We recall that a *division ring* T is a ring with identity in which the equations $xa = 1$, $ax = 1$ are soluble for each nonzero a in T. If the multiplication is commutative, then T is simply a field. We shall use the term "division ring" in the noncommutative case only. The simplest example of a division ring is the division ring of real quaternions; this is a 4-dimensional vector space over R with basis $1, i, j, k$ and multiplication table $i^2 = j^2 = k^2 = -1$, $ij = -ji = k$, $ik = -ki = j$, $jk = -kj = i$. Here $\mathbb{R}$ can be replaced by any of its subfields, by $\mathbb{Q}$ for example. The division ring of quaternions over $\mathbb{R}$ plays a significant role in geometry and analysis; division rings over algebraic number fields are important in number theory and in the arithmetic theory of algebraic groups. In general ring theory, the division rings acts as the "bricks" making up many of the objects occurring in the theory. Clearly, the centre of a division ring is a field, and the division ring is a vector space over it. If the dimension is finite, the division ring is said to be finite-dimensional (over the centre), and infinite-dimensional otherwise.

Investigations on linear groups include the one-dimensional case; the multiplicative group of a division ring T is just $\mathrm{GL}_1(T)$, and there are many articles devoted to it. As a rule, results about multiplicative groups of division rings are to be found in surveys on ring theory or division ring theory. Quite a lot is known about classical groups over division rings (Dieudonné (1971)). The most important result of recent years is Platonov's solution of an old problem of Tannaka and Artin, which we mentioned in 1.6.8. There is more detail on this and other papers on the same theme in the survey Zalesskij (1981).

The standard technique involving the action of the linear group on a vector space goes over in large measure to vector spaces over division rings. In particular, the concepts of irreducibility, complete reducibility and imprimitivity carry over without change to linear groups over division rings. Clifford's theorem (see 2.2.3) and Maschke's Theorem (4.2) continue to hold. The general elementary techniques are developed in Jacobson's book "Theory of Rings" (1943). One of the most frequently used facts is that a matrix ring of degree n over a division ring T cannot possess more than n pairwise orthogonal idempotents (that is, matrices e_j such that $e_j^2 = e_j$ and $e_j e_k = e_k e_j = 0$, for $k, j = 1, \ldots, k \neq j$). It follows that $\mathrm{GL}_n(T)$ has no elementary abelian subgroup of order p^{n+1} if p is prime different from the characteristic of T. In other words, the ranks of the abelian p-subgroups are bounded by n. This assertion is the basis of many general theorems about the structure of various classes of linear groups over division rings.

Quite a lot is known about the structure of finite linear groups over division rings. This question is intimately connected with the theory of the Schur index, one of the important aspects of the representation theory of finite groups. Let F be a subfield of $\mathbb{C}$ and ϕ an irreducible representation of a finite group G over $\mathbb{C}$. We consider the linear space R spanned by $\phi(G)$ over F. The irreducibility of ϕ gives that R is a simple algebra of finite dimension over F. By Wedderburn's theorem, R is isomorphic to a complete matrix algebra of some degree over a suitable division ring T. The dimension of T over its centre Z is finite, and it is known to be a square: $\dim T = d^2$, $d \in \mathbb{N}$. The number d is known as the *Schur index* of ϕ over F. We stress that the Schur index depends crucially on F; for example, the matrices of ϕ can be viewed as matrices over any superfield F_1 of F, and the Schur index of ϕ can be smaller over F_1 than over F. In general, the Schur index over F_1 is a divisor of that over F. If no reference is made to the field, it is usual to assume that $F = \mathbb{Q}$. The computation of the Schur index is one of the central problems in the representation theory of finite groups, and there is a large body of publications on this topic. The general theory is expounded in the book of Curtis and Reiner (1962); in particular, see Berman's supplement to the Russian translation. A good deal is known about the Schur indices of representations of the finite simple groups. Using results of this type, and the classification of the finite simple groups, Hartley and Shahabi (1982) have obtained the following theorem, which is a natural analogue of Jordan's classical result about finite linear groups over fields.

Theorem A. *There exists a function $h(n)$ taking positive integer values such that every finite linear group G of degree n over a division ring of characteristic* 0 *contains a metabelian normal subgroup H whose index $|G:H|$ does not exceed $h(n)$.*

Note that the theorem is false if the word "metabelian" is replaced by "abelian". A complete description of the finite linear groups over division rings exists only for $n = 1$ and 2.

Let T be a division ring of finite dimension k over its centre Z, so that $\mathrm{GL}_n(T)$ is embedded in $\mathrm{GL}_{nk}(Z)$. Thus, many questions about subgroups of $\mathrm{GL}_n(T)$ can be solved by viewing them as subgroups of $\mathrm{GL}_{nk}(Z)$. What can be said when k is infinite? The locally finite matrix groups over division rings are fairly amenable to investigation. However, no acceptably general uniformities for arbitrary division rings have been detected outside this class of groups. The following question is one of the most mysterious: is every p-group of matrices over a division ring locally finite? If p is the characteristic of the field, every p-group of matrices is unipotent. Therefore, the question is connected with the following one, where the positive answer is very probable:

Is every unipotent matrix group over a division ring locally nilpotent?

This problem has easy solutions for matrices of degrees 2 or 3. Serezhkin (1985) solved it for $n = 4$, while it is still open even for $n = 5$. However, the answer is "yes" for $p > (n - 1)(n/2 + 1)$ and $p = 0$; this follows from deep results of Heineken (1962). For more comments, see Zalesskij (1981), § 9.

Practically nothing else was known until the appearance in 1984 of a surprising result of Lichtman. He discovered that both of these questions have positive answers provided that the underlying division ring T is generated either by a polycyclic subgroup of $\mathrm{GL}_L(T) = T - \{0\}$ or by a finite-dimensional Lie subalgebra L of T (what this means is: T is the smallest division subring containing the centre Z of T, and P or L respectively). Later, Lichtman and Wehrfritz (1989) extended this result to "transcendental" division rings T. This term means that $T \otimes_Z D$ is a domain for every finite-dimensional division Z-algebra D. Many interesting division rings are transcendental. Wehrfritz (1989) also showed that a finitely generated subring of a division ring of characteristic 0 generated by a finite-dimensional Lie algebra is a residually finite ring. This enabled him to extend the residual method (see 2.5 above) to division rings of this type. Lichtman's original approach is also based on developing a residual method for noncommutative division rings. The statement of his basic residual theorem is more complicated than that of Wehrfritz; however, it includes division rings of quotients of group rings of polycyclic groups. Lichtman (1987) showed that a linear group G over such a division ring of characteristic 0 either contains a free subgroup of rank 2 or is soluble-by-finite. He has also estimated the derived length of G when G is soluble.

A good guide to this topic is the book of Wehrfritz and Shirvani (1986).

6.3. The Geometry of Linear Groups. The theory of linear groups was developed to a large extent because of the stimulus of geometrical applications. This is especially so for the classical groups, whose origin is essentially geometrical. Currently, the old connections between geometry and the classical groups are considered in the context of the general theory of algebraic transformation groups; thus, the place of geometry is taken by algebraic geometry, and instead of classical groups one has algebraic linear groups. The problem that is easiest to pose is that of describing the natural action of an algebraic linear group on the underlying vector space (or the adjoint action on the space of matrices). This topic has close connections with invariant theory. Unfortunately, it is not possible for us to discuss this circle of questions. We mention two facts as examples. Elashvili (1972) has determined all the irreducible algebraic groups in which the stabiliser of every vector is infinite, while Kac (1980) has determined the groups that have finitely many orbits; in both cases, the ground field is algebraically closed and of characteristic zero.

It would be hopeless to attempt to trace here the connections between the theory of linear groups and geometry. Some questions are discussed in § 9 of Zalesskij (1983). There are important geometrical applications of the theory of discrete groups generated by reflections (see Bourbaki (1971–72, 1968, 1975)); part of this theory relates to linear groups.

There is independent interest attaching to problems about the action of linear groups over finite fields on the natural vector spaces. This topic is closely connected with the general theory of finite transformation or permutation groups; for example, the problem of describing doubly transitive permutation groups involves that of describing the linear groups acting transitively on the nonzero vectors (see 3.4.6).

6.4. Generalisations of Linear Groups. The group of invertible infinite matrices of the form $E + m$, where E is the identity matrix and m a matrix with only finitely many nonzero entries, can be viewed as an infinite-dimensional analogue of the general linear group $\mathrm{GL}_n(R)$ over a ring R. This group is called a stable linear group; we denote it by $\mathrm{GL}^0_\infty(R)$, and it is important in algebraic K-theory. Bass's book (1968) pays very special attention to it; in particular, the normal subgroups are described there. One could adopt a different approach. Let M be any set of indices (not necessarily countable), and R' the ring anti-isomorphic to R. Let W be the free R'-module with basis $\{w_\mu\}_{\mu \in M}$ and $\mathrm{Aut}_{R'}(W)$ its automorphism group. We write $\mathrm{GL}^0_{|M|}(R)$ for the subgroup consisting of the elements g in $\mathrm{Aut}_{R'}(W)$ effectively permuting finitely many of the w_μ. If M is countably infinite, then $\mathrm{GL}^0_\infty(R)$ and $\mathrm{GL}^0_{|M|}(R)$ are isomorphic (note that the introduction of R' is a purely technical device; if, contrary to normal practice, we consider right modules, so that the product of an element w of W and an element r of R as wr, then R' becomes superfluous). The group $\mathrm{GL}^0_{|M|}(R)$ is not a normal subgroup of $\mathrm{Aut}_{R'}(W)$; its normal closure consists of the elements g of $\mathrm{Aut}_{R'}(W)$ such that $(E - g)W$ lies in the submodule spanned by finitely many of the w_μ. In fact, this subgroup seems to us to be the most natural analogue of $\mathrm{GL}_n(R)$. We shall denote it by $\mathrm{GL}_{|M|}(R)$. If the ring has an anti-automorphism, one can define infinite-dimensional analogues of the orthogonal, symplectic and unitary groups. Little is known about the factor-group $\mathrm{Aut}_{R'}(W)/\mathrm{GL}_{|M|}(R)$, even when R is a field. What are the abstract properties of subgroups of $\mathrm{GL}_{|M|}(R)$? Despite the fact that it is "infinite-dimensional", it is far from being the case that every abstract group can be embedded in $\mathrm{GL}_{|M|}(R)$. If G is a soluble subgroup of $\mathrm{GL}_{|M|}(R)$ with R a field, and U is the unipotent radical of G, then G/R turns out to be embeddable in a direct product of (infinitely many) finite groups (Zalesskij (1969)). This is a very strong restriction: for example, Borel subgroups of infinite

Chevalley groups over a field of characteristic $p \neq \operatorname{char} R$ do not satisfy this condition, and hence they cannot be embedded in $\mathrm{GL}_{|M|}(R)$. J. Hall (1988) has shown that the only infinite simple locally finite groups contained in $\mathrm{GL}_{|M|}(R)$, R a field of characteristic 0, are finitary alternating groups. The situation is more complicated when R is of prime characteristic. Cameron (1990) has described the subgroup of $\mathrm{GL}_{|M|}(R)$ generated by transvection groups (that is, groups conjugate to $\operatorname{diag}\left(E, \begin{pmatrix} 1 & R \\ 0 & 1 \end{pmatrix}\right)$, where E is the infinite identity matrix). Kosman (1986–7) has described the maximal unipotent subgroups in the case where R is a finite field.

We note that a natural group-theoretical generalization of $\mathrm{GL}_{|M|}(R)$ are the groups of finitary automorphisms of abstract groups; these are automorphisms whose fixed-point subgroups are of finite index (Zalesskij, 1975).

For every ring R, $\mathrm{GL}_n(R)$ is the group of invertible elements in the matrix ring $M_n(R)$. From this point of view, linear groups can be considered to be special examples of subgroups of the groups of invertible elements of rings. Experience shows that this view is not without importance if we are looking at a ring K generated by its invertible elements and having not too many two-sided ideals (for example, if it is simple). A fairly natural and very general analogue of rings like $M_n(R)$ is a ring K containing n pairwise orthogonal idempotents e_i (that is, elements such that $e_i^2 = e_i \neq 0$, $e_i e_j = e_j e_i = 0$ for $1 \leqslant i, j \leqslant n$, $i \neq j$), with the additional property that $1 = e_1 + \cdots + e_n$. If $K = M_n(R)$, for e_i we can take the matrix with zero in positions other than (i, j), and 1 in that position. Then K is the sum of the additive groups $e_i K e_j$, which serve as analogues of the matrices with zero everywhere except the (i, j)-position. There is some work concerned with extending results about normal subgroups of groups like $\mathrm{GL}_n(R)$ to subgroups of the groups of invertible elements of rings containing systems of orthogonal idempotents of a certain sort.

Another natural generalization of linear groups over rings are the automorphism groups of finitely generated modules over rings of various sorts. The endomorphism rings of such modules are very intensively studied within general ring theory. There are some papers devoted to the problem of when a finitely generated module is determined by its automorphism group. Very little is known about the structure of proper subgroups of full automorphism groups of finitely generated modules; we refer the reader to a series of papers by Wehrfritz (1975–1978), where certain facts about the structure of finitely generated linear groups over fields is carried over to finitely generated groups of automorphisms of finitely generated modules over commutative rings. The reader can also consult the survey of Markov, Mikhalev, Skornyakov and Tuganbaev on endomorphism rings of modules (1883).

References*

The monographs by Jordan (1870), Dickson (1901), Blichfeldt (1917), van der Waerden (1935), Dieudonné (1948), Suprunenko (1972), Wehrfritz (1973), Dixon (1971), Suprunenko (1958), and Wehrfritz and Shirvani (1986) are devoted specifically to the theory of linear groups. Of these, Jordan (1870), Dickson (1901), van der Waerden (1935) and Dieudonné (1948) are mainly of historical interest. Blichfeldt (1917) contains some results about finite linear groups that are still of interest to this day and are difficult to find anywhere else. Books devoted to classical groups are Artin (1957), Dieudonné (1948) and Dieudonné (1971). The contemporary level of study of subgroups of finite classical groups is reflected in Kleidman and Liebeck (1990). The small book of Dixon (1971) is a good introduction to the theory of linear groups. Infinite linear groups are treated in Wehrfritz (1973), largely from the viewpoint of general group theory. Work on soluble and nilpotent linear groups is covered very fully in Suprunenko (1972), Suprunenko (1958), Wehrfritz (1973). Zalesskij (1981) is an attempt to give an analytical survey of the ideas and methodology of the theory of linear groups. A complete bibliography of the previous 25 years can be found in Vol'vachev and Suprunenko (1965), Merzlyakov (1971 and 1978) and Zalesskij (1983); the aim there is a systematic description of the results obtained in that period.

Algebraic linear groups can be viewed as a natural generalisation of the classical groups. This is now an independent discipline in algebra; however, without it, a treatment of the contemporary state of the theory of linear groups would be impossible. Systematic exposition of the theory of algebraic groups is to be found in the monographs of Borel (1969) and Humphreys (1975). Introductions to the general theory are Borel (1966), the Seminar on Algebraic Groups and Related Subgroups (1970), and Kaplansky (1957); to representation theory, Humphreys (1976) and the Seminar on Algebraic Groups and Related Subgroups (1970); to the arithmetic theory, Humphreys (1980), Platonov and Rapinchuk (1991). The modern content of the theory of algebraic groups is surveyed in Platonov (1974), Platonov (1982), Platonov and Rapinchuk (1983) and the Seminar on Algebraic Groups and Related Subgroups (1970). An important part of the theory of algebraic groups is that dealing with the finite Chevalley groups and their representations. The main sources here are Steinberg (1967), the Seminar on Algebraic Groups and Related Subgroups (1970), Carter (1972 and 1985), Kostrikin and Chubarov (1985), and Lusztig (1984).

There is a close connection between finite linear groups and the representation theory of finite groups; the most important sources are Feit (1970), Kostrikin and Chubarov (1985), Kondrat'ev (1986), Curtis and Reiner (1962 and 1981–87). Another valuable reference work is Conway, Curtis, Norton, Parker and Wilson (1972). Lie groups also have a direct connection with linear groups; see Feudenthal and de Vries (1969), Zhelobenko and Shtern (1983), Ragunathan (1972), and Serre (1965) on this topic.

A significant part of the theory of linear groups over rings is expounded in books on algebraic K-theory; see Milnor (1971) and Bass (1968).

The remaining items in the reference list are either wide-spectrum monographs or books dealing with specialized topics that are important in the theory.

Artin, E. (1957): Geometric Algebra. Interscience, New York, Zbl.77,21

Bass, H. (1968): Algebraic K-Theory. Benjamin, New York, Zbl.174,303

Blichfeldt, H.R. (1917): Finite Collineation Groups. Univ. Chicago Press, Chicago, Jbuch.46,188

Borel, A. (1966): Linear Algebraic Groups. In: Algebraic Groups and Discontinuous Subgroups, 3–19. Am. Math. Soc., Providence, Zbl.205,505

Borel, A. (1969): Linear Algebraic Groups. Benjamin, New York, Zbl.186,332

* For the convenience of the reader, references to reviews in Zentralblatt für Mathematick (Zbl.), compiled using the MATH database, and Jahrbuch über die Fortschritte der Mathematik (Jbuch) have, as far as possible, been included in this bibliography.

Borel, A., Tits, J. (1965): Groupes réductifs. Publ. Math., Inst. Hautes Étud. Sci. *27*, 659–755, Zbl.145,174

Bourbaki, N. (1958): Algebra. Hermann, Paris, Zbl.102,272

Bourbaki, N. (1961–65): Algèbre commutative. Hermann, Paris, Zbl.108,40, Zbl.119,36, Zbl.205,343, Zbl.141,35

Bourbaki, N. (1971–72, 1968, 1975): Groupes et algèbres de Lie. Hermann, Paris 1971–72 (Ch. I, II, III), Zbl.213,41, Zbl.244.22007; 1968 (Ch. IV, V, VI), Zbl,186,330; 1975 (Ch. VII, VIII), Zbl.329.17002

Carter, R.W. (1972): Simple Groups of Lie Type. Wiley Interscience, Chichester, Zbl.248.20015

Carter, R.W. (1985): Finite Groups of Lie Type: Conjugacy Classes and Complex Characters. Wiley Interscience, Chichester, Zbl,567.20023

Conway, J.H., Curtis, R.T., Norton, S.P., Parker, R.A., Wilson, R.A. (1985): An Atlas of Finite Groups. Clarendon Press, Oxford, Zbl568,20001

Coxeter, H.S.M., Moser, W.O.J. (1972): Generators and Relations for Discrete Groups. Springer-Verlag, Berlin Heidelberg New York, Zbl.239,20040

Curtis, C.W., Reiner, I. (1962): Representation Theory of Finite Groups and Associative Algebras. Interscience Publ., New York, Zbl.131,256

Curtis, C.W., Reiner, I. (1981): Methods of Representation Theory with Applications to Finite Groups and Orders, Vol. 1. Wiley, New York, Zbl.469, 20001; Vol. 2 (1987)

Dickson, L.E. (1901): Linear Groups with an Exposition of the Galois Field Theory. Teubner, Leipzig, Jbuch.32,128

Dieudonné, J. (1948): Sur les groupes classiques. Hermann, Paris, Zbl.37,13

Dieudonné, J. (1971): La géometrie des groupes classiques. Troisième édition. Springer-Verlag, Berlin Heidelberg New York, Zbl.221,20056

Dixon, J.D. (1971): The Structure of Linear Groups. Van Nostrand, London, Zbl.232.20079

Dixon, J.D., du Sautoy, M.P.F., Mann, A., Segal, D. (1991): Analytic pro-*p*-groups. Lond. Math. Soc. Lect. Note Ser. *157*. Cambridge Univ. Press, Cambridge, Zbl.744.20001

Feit, W. (1970): The current situation in the theory of finite simple groups. Actes Congr. Int. Math., Vol. I, 55–93 (1971), Zbl.344,20008

Freudenthal, H., de Vries, H. (1969): Linear Lie Groups. Academic Press, New York, Zbl.377,22001

Humphreys, J.E. (1976): Ordinary and Modular Representations of Chevalley Groups. Springer-Verlag, Berlin Heidelberg New York, Zbl.341,20037

Humphreys, J.E. (1975): Linear Algebraic Groups. Springer-Verlag, Berlin Heidelberg New York, Zbl.325.20039

Humphreys, J.E. (1980): Arithmetic groups. Springer-Verlag, Berlin Heidelberg New York, Zbl.426.20029

Jordan, C. (1870): Traité des substitutions et des équations algébriques. Gauthier-Villars, Paris, Jbuch.3,42, Reprint: 1957 Blanchard, Paris, Zbl.78,12

Kaplansky, I. (1957): An Introduction to Differential Algebra. Hermann, Paris, Zbl.83,33

Kleidman, R., Liebeck, M.W. (1990): The subgroup structure of finite classical groups. Lond. Math. Soc. Lect. Note Ser. *129*. Cambridge Univ. Press, Cambridge, Zbl.697,20004

Kondrat'ev, A.S. (1986): Subgroups of finite Chevalley groups Usp. Mat. Nauk *41*, No. 1 (247), 57–96. English transl.: Russ. Math. Surv. *41*, No. 1, 65–118, Zbl,602.20041

Kostrikin, A.I., Chubarov, I.A. (1985): Representations of finite groups. Itogi Nauki Tekh., Ser. Algebra, Topologiya, Geom. *23*, 119–195, Zbl.606.20008. English transl.: J. Sov. Math. *40*, No. 3, 331–383 (1988)

Lusztig, G. (1984): Characters of reductive groups over a finite field. Ann. Math. Stud. *107*, Zbl.556,20033

Merzlyakov, Yu.I. (1971, 1978): Linear groups. Itogi Nauki Tekh., Ser. Algebra, Topologiya, Geom.; 1971, 75–110; 1978, 35–89, Zbl.324,20047, Zbl.401.20042. English transl.: J. Sov. Math. *1*, 571–593 (1973); *14*, 887–921 (1980)

Milnor, J. (1971): Introduction to Algebraic *K*-Theory. Princeton Univ. Press, Princeton, Zbl.237,18005

Platonov, V.P. (1966): The theory of algebraic linear groups and periodic groups. Izv. Akad. Nauk SSSR, Ser. Mat. *30*, 573–620, Zbl.146,44. English transl.: Transl., II. Ser., Am. Math. Soc. *69*, 61–110 (1968)

Platonov, V.P. (1974): Algebraic groups. Itogi Nauki Tekh., Ser. Algebra, Topologiya, Geom. 5–36, Zbl.295.20050. English transl.: J. Sov. Math. *4*, 463–482 (1976)

Platonov, V.P. (1982): Arithmetic theory of algebraic groups. Usp. Mat. Nauk *37*, No. 3 (225), 3–54, Zbl.502.20025. English transl.: Russ. Math. Surv. *37*, No. 3, 1–62

Platonov, V.P., Rapinchuk, A.S. (1983): Algebraic groups. Itogi Nauki Tekh., Ser. Algebra, Topologiya Geom. *21*, 80–134, Zbl.564.20023. English transl.: J. Sov. Math. *31*, 2939–2973 (1985)

Platonov, V.P., Rapinchuk, A.S. (1991): Algebraic Groups and Number Theory. Nauka, Moscow (Russian), Zbl.732.20027

Ragunathan, M.S. (1972): Discrete Subgroups of Lie Groups. Springer-Verlag, Berlin Heidelberg New York, Zbl.254.22005

Seminar on Algebraic Groups and Related Subgroups. Springer-Verlag, Berlin Heidelberg New York 1970, Zbl.192,362

Serre, J-P. (1964): Cohomologie Galoisienne. Springer-Verlag, Berlin Heidelberg New York, Zbl.128,263

Serre, J-P. (1965): Lie Algebras and Lie Groups. Benjamin, New York, Zbl.132,278

Serre, J-P. (1977): Arbres, amalgames, SL_2. Astérisque *46*, Zbl.369.20013. English transl.: Trees. Springer-Verlag, Berlin Heidelberg New York 1980

Shafarevich, I.R. (1985): Algebra I. Basic Notions of Algebra. Itogi Nauki Tekh., Ser. Sovrem. Probl. Mat., Fundam. Napravleniya *11*. English transl.: Encycl. Math. Sci. *11*, Springer-Verlag, Berlin Heidelberg New York 1990, Zbl.711.16001

Steinberg, R. (1967): Lectures on Chevalley Groups. Yale Univ. Press, Yale, Zbl.307.22001

Suprunenko, D.A. (1958): Soluble and Nilpotent Linear Groups. Belorussian Univ. Press, Minsk, Zbl.98,22. English transl.: Am. Math. Soc., Providence R.I. 1963

Suprunenko, D.A. (1972): Matrix Groups. Nauka, Moscow, Zbl.253.20074. English transl.: Am. Math. Soc., Providence R.I. 1976

Suprunenko, D.A., Tyshkevich, R.T. (1966): Permuting Matrices. Nauka Tekhnika, Minsk (Russian), Zbl.142,281. English transl.: Permutative matrices, Academic Press, New York 1968

Tits, J. (1972): Free subgroups in linear groups. J. Algebra *20*, 250–270, Zbl.236.20032

Vinberg, E.B., Shvartsman, O.V. (1988): Discrete Groups of Motions of Spaces of Constant Curvature. Itogi Nauki Tekh., Ser. Sovrem. Probl. Mat., Fundam. Napravleniya *29*, Geometry II, 147–259, Zbl.699.22017. English transl. in: Encycl. Math. Sc. *29*, Springer-Verlag, Berlin Heidelberg New York 1993

Vol'vachev, R.T., Suprunenko, D.A. (1967): Linear Groups. Itogi Nauki Tekhn., Ser. Algebra, Topologiya, Geom. 1965, 45–61. English transl.: Prog. Math. *5*, 39–56 (1969), Zbl.206,313

van der Waerden, B.L. (1935): Gruppen von Linearen Transformationen. Springer-Verlag, Berlin Heidelberg New York, Zbl.11,101

van der Waerden, B.L. (1971, 1967): Algebra I. Achte Auflage der Modernen Algebra. Springer-Verlag, Berlin Heidelberg New York 1971. Algebra II. Fünfte Auflage. Springer-Verlag, Berlin Heidelberg New York 1967, Zbl.221.12001; Zbl.192,330

Wehrfritz, B.A.F. (1973): Infinite Linear Groups. Springer-Verlag, Berlin Heidelberg New York, Zbl.261.20038

Wehrfritz, B.A.F., Shirvani, M. (1986): Skew Linear Groups. Cambridge Univ. Press, Cambridge, Zbl.602.20046

Wolf, J. (1967): Spaces of Constant Curvature. McGraw-Hill Book Comp., New York, Zbl.162,533

Zalesskij, A.E. (1981): Linear groups. Usp. Mat. Nauk *36*, No. 5 (221), 57–107, Zbl.475.20029. English transl.: Russ. Math. Surv. *36*, No. 5, 63–128 (1981)

Zalesskij, A.E. (1983): Linear groups. Itogi Nauki Tekh., Ser. Algebra, Topologiya, Geom. *21*, 135–182, Zbl.561.20033. English transl.: J. Sov. Math. *31*, 2974–3004 (1985)

Zhelobenko, D.P., Shtern, A.I. (1983): Representations of Lie Groups. Nauka, Moscow (Russian), Zbl.521.22006

Author Index

Subject Index

Encyclopaedia of Mathematical Sciences

Editor-in-Chief: R. V. Gamkrelidze

Geometry

Volume 28: **R.V. Gamkrelidze** (Ed.)

Geometry I

Basic Ideas and Concepts of Differential Geometry

1991. VII, 264 pp. 62 figs. ISBN 3-540-51999-8

Volume 29: **E.B. Vinberg** (Ed.)

Geometry II

Geometry of Spaces of Constant Curvature

1993. Approx. 250 pp. 87 figs. ISBN 3-540-52000-7

Volume 48: **Yu. D. Burago, V. A. Zalgaller** (Eds.)

Geometry III

Theory of Surfaces

1992. VIII, 256 pp. ISBN 3-540-53377-X

Volume 70: **Yu. G. Reshetnyak** (Ed.)

Geometry IV

Nonregular Riemannian Geometry

1993. Approx. 270 pp. 58 figs. ISBN 3-540-54701-0

Algebraic Geometry

Volume 23: **I. R. Shafarevich** (Ed.)

Algebraic Geometry I

Algebraic Curves. Algebraic Manifolds and Schemes

1993. Approx. 330 pp. 9 figs. ISBN 3-540-51995-5

Volume 35: **I. R. Shafarevich** (Ed.)

Algebraic Geometry II

Cohomological Methods in Algebra. Geometric Applications to Algebraic Surfaces

1995. Approx. 270 pp. ISBN 3-540-54680-4

Volume 36: **A. N. Parshin, I. R. Shafarevich** (Eds.)

Algebraic Geometry III

Complex Algebraic Manifolds

1995. Approx. 270 pp. ISBN 3-540-54681-2

Volume 55: **A. N. Parshin, I. R. Shafarevich** (Eds.)

Algebraic Geometry IV

Linear Algebraic Groups. Invariant Theory

1993. Approx. 300 pp. 9 figs. ISBN 3-540-54682-0

Number Theory

Volume 49: **A. N. Parshin, I. R. Shafarevich** (Eds.)

Number Theory I

Fundamental Problems, Ideas and Theories

1994. Approx. 340 pp. 16 figs.
ISBN 3-540-53384-2

Volume 62: **A. N. Parshin, I. R. Shafarevich** (Eds.)

Number Theory II

Algebraic Number Theory

1992. VI, 269 pp. ISBN 3-540-53386-9

Volume 60: **S. Lang**

Number Theory III

Diophantine Geometry

1991. XIII, 296 pp. 1 fig. ISBN 3-540-53004-5